Drug Metabolism and Pharmacokinetics Quick Guide

Siamak Cyrus Khojasteh
Harvey Wong · Cornelis E.C.A. Hop

Drug Metabolism and Pharmacokinetics Quick Guide

 Springer

Siamak Cyrus Khojasteh
khojasteh.cyrus@gene.com

Harvey Wong
wong.harvey@gene.com

Cornelis E.C.A. Hop
hop.cornelis@gene.com

Genentech, Inc.
1 DNA Way
San Francisco, California
USA

ISBN 978-1-4419-5628-6   e-ISBN 978-1-4419-5629-3
DOI 10.1007/978-1-4419-5629-3
Springer New York Dordrecht Heidelberg London

Library of Congress Control Number: 2011923538

© Springer Science+Business Media, LLC 2011
All rights reserved. This work may not be translated or copied in whole or in part without the written permission of the publisher (Springer Science +Business Media, LLC, 233 Spring Street, New York, NY 10013, USA), except for brief excerpts in connection with reviews or scholarly analysis. Use in connection with any form of information storage and retrieval, electronic adaptation, computer software, or by similar or dissimilar methodology now known or hereafter developed is forbidden.
The use in this publication of trade names, trademarks, service marks, and similar terms, even if they are not identified as such, is not to be taken as an expression of opinion as to whether or not they are subject to proprietary rights.
While the advice and information in this book are believed to be true and accurate at the date of going to press, neither the authors nor the editors nor the publisher can accept any legal responsibility for any errors or omissions that may be made. The publisher makes no warranty, express or implied, with respect to the material contained herein

Printed on acid-free paper

Springer is part of Springer Science+Business Media (www.springer.com)

This book is dedicated to:

Zarrin, Sohrob and my parents, Maryam and Mahmoud
Sally, Ethan and Matthew
Giti, Patrick and Chloe

For their love and support

Preface

Drug discovery is complicated and yet rewarding in many ways. We have come to appreciate that drug discovery is possible when we are constantly making important and timely decisions for synthesis of superior compounds that have the potential for being a safe and an effective drug. Drug metabolism and pharmacokinetics play an integral role in this process.

Drug Metabolism and Pharmacokinetics Quick Guide is intended for broad readership for those working or interested in drug discovery from various disciplines such as medicinal chemistry, pharmacology, drug metabolism and pharmacokinetics, bioanalysis, clinical sciences, biochemistry, pharmaceutics and toxicology. It provides, for the first time, a completely integrated look at multiple aspects of ADME sciences (absorption, distribution, metabolism, and excretion) in a summary format that is clear, concise, and self-explanatory. We have minimized the amount of prior knowledge required for the reader by providing the basics of each concept. This reference book is meant to be used day to day and provides many useful tables (used for data interpretation), figures and factoids. The factoids are intended to be short and relevant to the topic discussed that would allow another dimension to the discussions.

Acknowledgments

To put this book together required the help of many. We like to acknowledge the editorial help of Ronitte Libedinsky and comments by Ignacio Aliagas, Jeff Blaney, John R. Cashman, Patrick Dansette, Xiao Ding, Peter W. Fan, Jason S. Halladay, James P. Driscoll, Jeffrey P. Jones, Jane R. Kenny, Walter A. Korfmacher, Lichuan Liu, Xingrong Liu, Anthony Y. H. Lu, Joseph Lubach, Dan Ortwine, K. Wayne Riggs, Young Shin, George R. Tonn, and Joseph A Ware.

Contents

About the Authors

Dr. Siamak Cyrus Khojasteh received his BS from University of California at Berkeley and his PhD from University of Washington in Medicinal Chemistry under the direction of Dr. Sidney D. Nelson. Dr. Khojasteh leads the Metabolism efforts at Genentech (South San Francisco) and leads a team of about 20 scientists and prior to that he was a Senior Research Scientist at Pfizer (Groton, CT). His research interest is the mechanism of biotransformation particularly with the formation of reactive metabolites by P450 or non-P450 enzymes.

Dr. Harvey Wong graduated from the University of British Columbia with a Ph.D. in Pharmacokinetics and Biopharmaceutics. Following graduation, he worked at the DuPont Pharmaceutics Company followed by Bristol-Myers Squibb in the area of Neuroscience Drug Discovery. Currently, Harvey is a Senior Scientist in the Department of Drug Metabolism and Pharmacokinetics at Genentech, Inc. working in the areas of oncology and immunology. He is involved in pharmacokinetic modeling and defining preclinical PK-PD relationships for drug candidates in both therapeutic areas. Harvey has published over 70 publications and abstracts.

Dr. Cornelis E.C.A. Hop is supervising the Small Molecule Drug Metabolism & Pharmacokinetics Department at Genentech (South San Francisco) and leads a team of about 55 scientists involved in acquisition and interpretation of ADME data in support of drug discovery and development. Before that he was a senior director at Pfizer (Groton, CT) and a Senior Research Fellow at Merck (Rahway, NJ). Dr. Hop has extensive experience in ADME sciences and biotransformation, PK prediction and bioanalysis in particular. He has authored more than 100 publications in refereed journals and several book chapters and made more than 50 external oral presentations at conferences and universities.

Chapter 1
Pharmacokinetics

Abstract
Pharmacokinetics serves as a "tool" that provides a quantitative description of what the body does to a compound/drug that is administered. Processes such as absorption, distribution, metabolism, and excretion are described by pharmacokinetic parameters. Characterization of pharmacokinetics allows for selection of appropriate and efficacious dosing regimens. In drug discovery and preclinical development, an understanding of pharmacokinetics in animals allows for selection and advancement of drug candidates. This chapter describes and defines basic pharmacokinetic parameters and their derivation. Empirical compartmental modeling and simple physiological modeling are described. In addition, useful tables for the interpretation of pharmacokinetic data are provided.

Contents

1.1 ABBREVIATIONS

α	Distribution phase rate constant for a 2-compartment model
A	y-Axis intercept for distribution phase
AUC	Area under the blood/plasma concentration-time profile

S.C. Khojasteh et al., *Drug Metabolism and Pharmacokinetics Quick Guide*, DOI 10.1007/978-1-4419-5629-3_1,
© Springer Science+Business Media, LLC 2011

$AUC_{Extravascular}$	AUC following an extravascular dose (i.e., oral, subcutaneous, intraperitoneal)
AUC_{IV}	AUC following an intravenous dose
β	Terminal elimination phase rate constant for 2-compartment model
B	y-Axis intercept for terminal elimination phase
C_{in}	Concentration entering liver or organ
$C_{initial}$	Initial concentration
C_{out}	Concentration exiting the liver or organ
CL	Clearance
CL_{int}	Intrinsic metabolic clearance
C_{max}	Highest or peak blood/plasma concentration
$C_{max}(ss)$	C_{max} at steady state
E	Extraction ratio
F	Bioavailability
f_u	Unbound fraction in blood/plasma
f_{uT}	Unbound fraction in tissue
k	First-order terminal elimination rate constant
k_{12}	Intercompartment rate constant from central to peripheral compartment
k_{21}	Intercompartment rate constant from peripheral to central compartment
k_a	First-order absorption rate constant
MRT	Mean residence time
Q	Hepatic blood flow
R	Accumulation ratio
T	Tau; dosing interval
$t_{1/2}$	Half-life
t_{max}	Time at which C_{max} occurs
V_β	Nomenclature for V_d for a 2-compartment model
V_C	Volume of the central compartment
V_d	Volume of distribution
V_p	Volume of plasma
V_{SS}	Volume of distribution at steady state
V_T	Volume of tissue

I.2 BASIC CONCEPTS

I.2.I Area Under the Concentration-Time Profile (AUC)

Area under the concentration-time profile (AUC) is an important measurement of exposure (see Fig. 1.1). Assessment of AUC tends to be a more "robust" measurement of drug exposure than the measurement of drug concentrations in blood/plasma at a single timepoint. Errors in drug concentration measurements at single timepoints generally have much less impact on AUC.

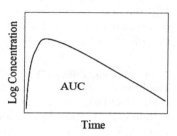

FIGURE 1.1. Area under the concentration-time profile (AUC).

AUC is expressed as units of concentration × time (e.g., µg × h/mL). Under linear conditions, AUC increases proportionally with dose.

> For a reasonable estimation of AUC, collection of plasma/blood samples should occur long enough such that the portion of the AUC that is extrapolated beyond the last measured concentration accounts for <30% of the AUC estimate.

1.2.2 Highest or Peak Blood/Plasma Concentration (C_{max})

Highest or peak blood/plasma concentration (C_{max}) is an important measurement of exposure (see Fig. 1.2).

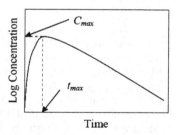

FIGURE 1.2. Highest or peak blood/plasma concentration (C_{max}) and time at which C_{max} occurs (t_{max}).

C_{max} is expressed in concentration units (i.e., µg/mL). Similar to AUC, C_{max} increases proportionally to dose under linear conditions. If performing noncompartmental analysis, this concentration is defined by the chosen sample collection times and the "true" C_{max} is not known. After intravenous dosing, C_{max} usually occurs immediately following the completion of dosing.

1.2.3 Time at Which C_{max} Occurs (t_{max})

The time at which C_{max} occurs is defined as the t_{max} (see Fig. 1.2). The t_{max} is expressed in units of time (i.e., hours).

t_{max} is a measurement of the rate of extravascular absorption into the systemic circulation. Similar to C_{max}, when performing noncompartmental analysis, the t_{max} is defined by the chosen sample collection times and the "true" t_{max} is not known.

1.2.4 Bioavailability (*F*)

The bioavailability (*F*) is the extent to which an extravascular (i.e., oral, subcutaneous, intraperitoneal, etc.) dose is available to the systemic circulation in relation to an intravenous dose. Bioavailability is expressed as a percentage. The calculation of bioavailability is shown in the following equation:

$$F = \frac{\text{AUC}_{\text{Extravascular}}/\text{Dose}_{\text{Extravascular}}}{\text{AUC}_{\text{IV}}/\text{Dose}_{\text{IV}}} \times 100 \qquad (1.1)$$

1.2.4.1 Sample Calculation

Oral dose – 2 mg/kg
Oral AUC – 20 μg × h/mL

$\longrightarrow F = \dfrac{20/2}{50/1} \times 100 = 20\%$

Intravenous dose – 1 mg/kg
Intravenous AUC – 50 μg ×h/mL

For oral dosing, a high bioavailability does not necessarily translate to great oral exposure. For compounds with very high clearance (CL), often you may observe high *F*% estimates due to a very low AUC$_{\text{IV}}$. In addition, pharmacokinetic study designs often use high oral doses when compared to the intravenous dose. Since an oral dose is absorbed into the portal vein and must pass through the liver prior to entering the blood, saturation of hepatic first pass metabolism may occur. This scenario can result in a high *F* value that on occasion may be >100%, especially when high oral doses are given. Oral exposure of compounds should be ranked based upon both *F*% and oral AUC and interpreted in the context of the oral dose given.

1.2.5 Clearance (CL)

CL is the volume of blood cleared of compound/drug per unit time. CL units are volume/time (i.e., mL/min). CLs by various organs are additive. The total body CL of a compound is a summation of the CL contributions of various organs.

$$CL_{total\ body} = CL_{liver} + CL_{renal} + CL_{other} \qquad (1.2)$$

CL can be calculated as follows:

$$CL = \frac{Dose}{AUC} \qquad (1.3)$$

$$CL = V_d \times k \qquad (1.4)$$

and

$$k = \frac{0.693}{t_{1/2}} \qquad (1.5)$$

where

k is the first-order elimination rate constant (units of time^{-1})
$t_{1/2}$ is the half-life (units of time)
V_d is the volume of distribution (units of volume)

$$CL = \frac{Elimination\ rate}{Concentration} \qquad (1.6)$$

Hepatic metabolism is the primary route of elimination for most compounds. A general assumption, although not always true, is that CL_{liver} (due to hepatic metabolism) is equal to $CL_{total\ body}$.

1.2.6 Volume of Distribution (V_d)

The volume of distribution is a proportionality constant relating half-life to CL. Volume of distribution units are volume terms (i.e., L). Volume of distribution is related to CL and half-life as follows:

$$V_d = CL \times \frac{t_{1/2}}{0.693} \qquad (1.7)$$

From a physiological point of view, the volume of distribution consists of the sum of the plasma volume and tissue volume as described in the following equation:

$$V_d = V_p + V_T\left(\frac{f_u}{f_{uT}}\right) \tag{1.8}$$

where

V_p is the volume of plasma
V_T is the volume of tissue
f_u is the unbound fraction in plasma
f_{uT} is the unbound fraction in tissue

The tissue volume is influenced by the degree of protein and tissue binding. For example, an increase in protein binding in the absence of tissue binding changes will cause a decrease in f_u and a decrease in the volume of distribution.

For a compound exhibiting one compartment characteristics

$$V_d = V_{SS} = V_C \tag{1.9}$$

where V_{SS} is the volume of distribution at steady state and V_C is the volume of the central compartment.

For compounds showing multiple compartment characteristics (see Fig. 1.3)

$$V_d(V_\beta \text{ for a 2 - compartment model}) > V_{SS} > V_C \tag{1.10}$$

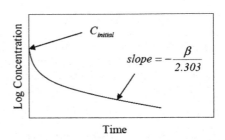

FIGURE 1.3. Concentration-time profile showing derivation of $C_{initial}$ and β.

$$V_C = \frac{\text{Dose}}{C_{initial}} \tag{1.11}$$

$$V_d = \frac{\text{Dose}}{\text{AUC} \times \beta} \tag{1.12}$$

It is important to note that CL and volume of distribution are independent terms. The half-life is dependent on both CL and volume of distribution.

1.2.7 Half-Life ($t_{1/2}$)

The time it takes for the amount of compound in the body to decrease to one half of the original amount is the half-life (see Fig. 1.4). Half-life is expressed in units of time (i.e., hours). Half-life can be calculated as follows:

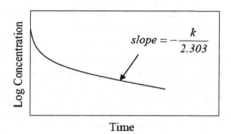

FIGURE 1.4. Concentration-time profile showing derivation of k.

$$t_{1/2} = \frac{0.693}{k} \qquad (1.13)$$

For a 2-compartment model (see Sect. 1.3)

$$t_{1/2} = \frac{0.693}{\beta} \qquad (1.14)$$

The degree of accumulation following multiple doses is governed by the half-life/elimination rate constant. The Accumulation ratio R is defined as follows:

$$R = \frac{C_{\max}(\text{ss})}{C_{\max}} \qquad (1.15)$$

or

$$R = \frac{1}{1 - e^{-kT}} \qquad (1.16)$$

where

C_{max} is the highest or peak blood/plasma concentration after a single dose

$C_{max}(ss)$ is C_{max} at steady state

T (tau) is the dosing interval in units of time (i.e., hours)

Half-life is related to volume of distribution and CL as follows:

$$t_{1/2} = \frac{V_d \times 0.693}{CL} \qquad (1.17)$$

An increase in volume of distribution or a decrease in CL will cause an increase in half-life.

> Estimation of the terminal half-life is sometimes influenced by the sensitivity of the bioanalytical assay. In cases where there is inadequate sensitivity and/or low doses are administered, the terminal phase is not adequately characterized resulting in a much shorter than actual half-life estimate.

1.2.8   Mean Residence Time (MRT)

Mean residence time (MRT) is the average time that molecules of a dose spend in the body or the mean transit time of molecules through the body. MRT is expressed in units of time (i.e., hours). MRT can be calculated as follows:

$$MRT = \frac{1}{k} \qquad (1.18)$$

1.3   COMPARTMENTAL MODELING

Compartmental modeling represents the body empirically as a number of compartments. Most drugs can be described adequately by either a one- or two-compartment model. In some cases, a three-compartment model is required. The general rule is to use the fewest number of compartments that can adequately describe the data. Basics of one- and two-compartment model will be presented here. For more detailed descriptions, please see references provided at the end of the chapter.

1.3.1   One-Compartment Model

Compounds exhibiting kinetics described by a one-compartment model have a log concentration-time profile that is monophasic in nature (see Fig. 1.5).

1.3.1.1 *Intravenous Dosing*
k – first-order elimination rate constant (units of time^{-1})

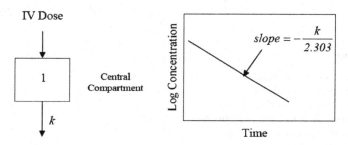

FIGURE 1.5. One-compartment model following a single intravenous dose.

Concentration following an intravenous dose can be calculated using the following equation:

$$C = C_{\text{initial}} e^{-kt} \qquad (1.19)$$

where

C is concentration
C_{initial} is the initial concentration following IV dosing
k is the first-order elimination rate constant
t is time

1.3.1.2 *Extravascular Dosing*
k – first-order elimination rate constant (units of time^{-1})
k_a – first-order absorption rate constant (units of time^{-1})

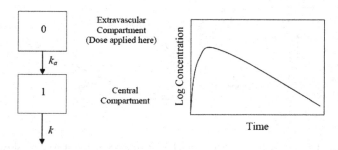

FIGURE 1.6. One-compartment model following a single extravascular dose.

Concentration following an extravascular dose (see Fig. 1.6) can be calculated using the following equation:

$$C = \frac{F \times \text{Dose}}{V_d} \frac{k_a}{(k_a - k)} (e^{-kt} - e^{-k_a t}) \qquad (1.20)$$

where

C is the concentration
F is bioavailability
V_d is volume of distribution
k_a is first-order absorption rate constant
k is first-order elimination rate constant
t is time

1.3.2 Two-Compartment Model

Compounds exhibiting kinetics described by a two-compartment model have a log concentration-time profile that is biphasic in nature (see Fig. 1.7).

1.3.2.1 Intravenous Dosing

k – first-order elimination rate constant (units of time^{-1})
k_{12} – intercompartment rate constant from central to peripheral compartment (units of time^{-1})
k_{21} – intercompartment rate constant from peripheral to central compartment (units of time^{-1})

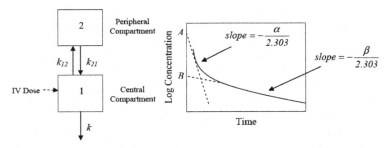

FIGURE 1.7. Two-compartment model following a single intravenous dose.

Concentration following an intravenous dose can be calculated using the following equation:

$$C = Ae^{-\alpha t} + Be^{-\beta t} \tag{1.21}$$

where

C is concentration
α is distribution rate constant
β is terminal elimination rate constant
A is y-axis intercept for distribution phase
B is y-axis intercept for elimination phase
t is time

Relationships of α, β, A, and B (macroconstants) to k_{12}, k_{21}, and k (microconstants) are as follows:

$$A = \frac{\text{Dose}(k_{21} - \alpha)}{V_C(\beta - \alpha)} \tag{1.22}$$

$$B = \frac{\text{Dose}(k_{21} - \beta)}{V_C(\alpha - \beta)} \tag{1.23}$$

$$\alpha + \beta = k_{12} + k_{21} + k \tag{1.24}$$

$$\alpha \times \beta = k_{21} \times k \tag{1.25}$$

where V_C is the volume of the central compartment.

Many of these parameters can be estimated using commercial pharmacokinetic software such as WinNonlin® and Kinetica®. For more detailed information about compartmental models, please refer to references at the end of this chapter.

1.4 PHYSIOLOGICAL MODELING

Physiological modeling differs from compartmental modeling in that parameters estimated have physiological meaning rather than just being empirically descriptive. As the liver is the main organ of elimination for most compounds, we describe below a simple physiological model of liver organ CL (see Fig. 1.8).

1.4.1 Hepatic Organ Clearance (Well-Stirred Model)

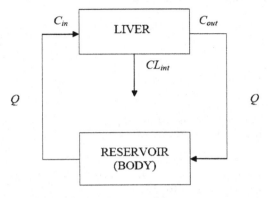

FIGURE 1.8. Simple physiological model of hepatic organ clearance.

C_{in} is the concentration entering liver or organ
C_{out} is the concentration exiting the liver or organ
Q is the hepatic blood flow
CL_{int} is the intrinsic metabolic CL

Hepatic CL is defined as:

$$CL_{hepatic} = Q \times E \tag{1.26}$$

$$E = \frac{(C_{in} - C_{out})}{C_{in}} \tag{1.27}$$

or

$$E = \frac{f_u \times CL_{int}}{f_u \times CL_{int} + Q} \tag{1.28}$$

where E is the extraction ratio and f_u is the unbound fraction.

The extraction ratio is the fraction of drug removed by the liver with each pass through the liver. For example, an E of 0.5 means that 50% of the drug passing through the liver is removed.

For drugs with HIGH intrinsic CL, hepatic CL is dependent on hepatic blood flow.

$$CL_{hepatic} = Q \frac{f_u \times CL_{int}}{f_u \times CL_{int} + Q} \xrightarrow{\text{approaches}} CL_{hepatic} \approx Q \tag{1.29}$$

For drugs with LOW intrinsic CL, hepatic CL is dependent on unbound fraction and intrinsic CL.

$$CL_{hepatic} = Q \frac{f_u \times CL_{int}}{f_u \times CL_{int} + Q} \xrightarrow{\text{approaches}} CL_{hepatic} \approx f_u \times CL_{int} \quad (1.30)$$

Since drugs are commonly cleared by the liver, hepatic CL often approximates total body CL and can be categorized based upon a percent of hepatic blood flow.

LOW clearance   <30% of hepatic blood flow
MODERATE clearance   30–70% of hepatic blood flow
HIGH clearance   >70% of hepatic blood flow

Examples of LOW, MODERATE, and HIGH clearance compounds are as follows:

LOW	MODERATE	HIGH
Carbamazepine	Aspirin	Alprenolol
Diazepam	Codeine	Cocaine
Ibuprofen	Cyclosporine	Meperidine
Nitrazepam	Ondansetrone	Morphine
Paroxetine	Nifedipine	Nicotine
Salicylic acid	Nortriptyline	Nitroglycerine
Warfarin		Propoxyphene
		Verapamil

Adapted from Rowland and Tozer (2011)

Clearance Estimate is Higher than Hepatic Blood Flow

Categorization of CL as low, moderate, and high based upon hepatic blood flow is important in the assessment of the pharmacokinetics of drugs and drug candidates. Most drugs are orally administered and compounds with low hepatic CL result in better $F\%$ and oral exposure. Sometimes certain compounds will show CLs higher than hepatic blood flow. Three possible physiological reasons for this phenomenon include:

1. Compound is cleared by extrahepatic elimination pathways. Although hepatic metabolism is the most common route of drug elimination, it is not the only route.
2. Plasma CL is often estimated rather than blood CL (blood bioanalytical assays are much more difficult to perform than plasma assays). For compounds that preferentially distribute into red blood cells, the estimate of total body CL will be overestimated by the plasma CL. This can be investigated by measuring the blood to plasma ratio of the compound of interest.

Continued

3. Compounds that are given intravenously and have extensive lung uptake can sometimes have CL estimates that are greater than the cardiac output.

Finally, compound degradation in blood or plasma can result in an overestimation of CL. In some cases, CL estimates for compounds cleared by liver may be higher than hepatic blood flow because of instability in plasma/blood.

1.5 PHYSIOLOGICAL PARAMETERS IN VARIOUS SPECIES

The following section contains tables with physiological parameters (blood flows (see Table 1.1) and volumes of various fluids and organs (see Table 1.2)), which can aid in the interpretation of CL and volume of distribution estimates.

TABLE 1.1. Blood flows to various organs and tissues for mouse, rat, dog, monkey, and human

Blood flows in mL/min (mL/min/kg – assumes listed body weight)					
Species (weight)	Mouse (0.02 kg)	Rat (0.25 kg)	Dog (10 kg)	Monkey (5 kg)	Human (70 kg)
Cardiac output	8.0 (400)	74 (296)	1,200 (120)	1,086 (217)	5,600 (80)
Glomerular filtration rate (GFR)	0.28 (14)	1.3 (5.2)	61.3 (6.13)	10.4 (2.08)	125 (1.79)
Tissues					
Adipose	–	0.4 (1.6)	35 (3.5)	20 (4.0)	260 (3.71)
Bone	–	13.5 (53.9)	–	–	218 (3.12)
Brain	0.46 (23)	1.3 (5.2)	45 (4.5)	72 (14.4)	700 (10.0)
Heart	0.28 (14)	3.9 (15.6)	54 (5.4)	60 (12.0)	240 (3.4)
Kidneys	1.30 (65)	9.2 (37)	216 (21.6)	138 (27.6)	1,240 (17.7)
Liver (total)	1.8 (90)	13.8 (55.2)	309 (30.9)	218 (43.6)	1,450 (20.7)
Hepatic artery	0.35 (18)	2.0 (8.0)	79 (7.9)	51 (10.2)	300 (4.29)
Portal vein	1.45 (73)	9.8 (39)	230 (23)	167 (33.4)	1,150 (16.4)
Lung	0.070 (3.5)	2.3 (9.3)	106 (10.6)	–	–
Muscle	0.91 (46)	7.5 (30)	250 (25)	90 (18.0)	750 (10.7)
Skin	0.41 (21)	5.8 (23)	100 (10)	54 (10.8)	300 (4.3)
Thyroid	–	–	–	–	83.2 (1.19)

Data from Davies and Morris (1993) and Brown et al. (1997)

TABLE 1.2. Volumes of various body fluids and organs in mouse, rat, dog, monkey, and human

Volume (mL)	Species (weight)				
	Mouse (0.02 kg)	Rat (0.25 kg)	Dog (10 kg)	Monkey (5 kg)	Human (70 kg)
Blood	1.7	13.5	900	367	5,200
Plasma	1.0	7.8	515	224	3,000
Total body water	14.5	167	6,036	3,465	42,000
Intracellular fluid	–	92.8	3,276	2,425	23,800
Extracellular fluid	–	74.2	2,760	1,040	18,200
Liver	1.3	19.6	480	135	1,690
Brain	–	1.2	72	–	1,450

Data from Davies and Morris (1993)

References

Brown RP, Delp MD, Lindstedt SL et al (1997) Physiological parameter values for physiologically based pharmacokinetic models. Toxicol Ind Health 13:407–484

Davies B, Morris T (1993) Physiological parameters in laboratory animals and humans. Pharm Res 10:1093–1095

Gilbaldi M, Perrier D (1982) Pharmacokinetics, 2nd edn. Marcel Dekker, New York

Rowland M, Tozer TN (2011) Clinical pharmacokinetics and pharmacodynamics: concepts and applications. Lippincott Williams & Wilkins, Baltimore

Shargel L, Yu ABC (1999) Applied biopharmaceutics & pharmacokinetics, 4th edn. Appleton & Lange, Stamford

Wagner JG (1993) Pharmacokinetics for the pharmaceutical scientist. Technomic Publishing Company, Lancaster

Additional Reading

Wilkinson GR, Shand DG (1975) Commentary: a physiological approach to hepatic drug clearance. Clin Pharmacol Ther 18:377–390

Chapter 2
Drug Metabolizing Enzymes

Abstract

Metabolism is the major elimination pathway of a drug from the body. Drug metabolizing enzymes (DMEs) are mainly present in the liver, intestine, and blood and are responsible for converting lipophilic drugs to more hydrophilic compounds to facilitate their excretion from the body. DMEs are classified as either Phase I or Phase II enzymes. Phase I DMEs are responsible for oxidation, reduction, and hydrolysis, and Phase II DME are responsible for conjugation (not necessarily sequential). Here we discuss the DMEs involved in Phase I and Phase II reactions, their subcellular locations, cofactors, organ distributions, mechanisms of reactions, and typical substrates and inhibitors.

Contents

2.1 ABBREVIATIONS

ABT	Aminobenzotriazole
ADH	Alcohol dehydrogenase
AKR	Aldo-keto reductase
ALDH	Aldehyde dehydrogenase
AO	Aldehyde oxidase
AZT	3'-Azido-3'-deoxythimidine
BSO	L-Buthionine-sulfoximine
CDNB	1-Chloro-2,4-dinitrobenzene

S.C. Khojasteh et al., *Drug Metabolism and Pharmacokinetics Quick Guide*, DOI 10.1007/978-1-4419-5629-3_2,
© Springer Science+Business Media, LLC 2011

CL	Clearance
DCNP	2,6-Dichloro-4-nitrophenol
DME	Drug metabolizing enzyme
EC	Enzyme classification number based on enzyme function
EH	Epoxide hydrolase
ER	Endoplasmic reticulum (i.e., microsomes)
FAD	Flavin adenine dinucleotide
FMO	Flavin-containing monooxygenase
GI	Gastrointestinal
GST	Glutathione S-transferase
LM	Liver microsomes
MAO	Monoamine oxidase
mCPBA	*m*-Chloroperoxybenzoic acid
NAC	*N*-Acetylcysteine
NAD	Nicotinamide adenine dinucleotide
NADPH	Nicotinamide adenine dinucleotide phosphate
NAT	*N*-Acetyltransferase
P450	Cytochrome P450
PAPS	3′-Phosphoadenosine-5′-phosphosulfate
PM	Poor metabolizer
NQO	NADPH:quinone reductase
SAM	S-Adenosyl methionine
SULT	Sulfotransferase
UDPGA	Uridine diphosphoglucuronic acid
UGT	Uridine diphosphate glucuronosyltransferase
XDH	Xanthine dehydrogenase
XO	Xanthine oxidase

2.2   BASIC CONCEPTS AND DEFINITIONS

Metabolism is the main elimination pathway of xenobiotics from the body (see Table 2.1).

In equation form, the total in vivo clearance (CL) is represented as:

$$CL_{total} = CL_{renal} + CL_{biliary} + CL_{metabolism} + CL_{others} \qquad (2.1)$$

CL_{renal} and $CL_{biliary}$ = unchanged drug being cleared from urine and bile, respectively.

$CL_{metabolism}$ = the contribution of metabolism, which includes both hepatic and extrahepatic metabolism.

CL_{other} = any unaccounted for CL
Note that CL terms are additive.

R. Tecwyn Williams first proposed in 1959 that drug meta-bolizing enzymes (DMEs) should be classified into two categories: Phase I and Phase II.

Phase I DMEs are involved in oxidation, reduction, or hydroly-sis (see Table 2.2). The reactions by these enzymes are also called "functionalization" reactions, but this term is not descriptive enough to cover the scope of this biotransformation pathway.

Phase II DMEs are involved in conjugation reactions (see Table 2.3).

> Drug transporters are called Phase III enzymes, and they facili-tate the entering and leaving of drugs in cells (see Chap. 4)

TABLE 2.1. Routes of elimination of marketed drugs (Williams et al. 2004)

Route of elimination	Percentage of marketed drugs
Metabolism	70% (50% P450, 12% UGT, 5% esterases, 3% others)[a]
Urine	20%
Bile	10%

[a]% Contribution of different enzymes based on 70% metabolism of a molecule

UGT Uridine diphosphate glucuronosyltransferase

TABLE 2.2. Phase I drug metabolizing enzymes (DMEs) and their subcellu-lar location or blood

Pathway	Enzyme (abbreviation) EC number	Subcellular location or blood
Oxidation	Alcohol dehydrogenase (ADH) 1.1.1	Cytosol, blood vessels
	Aldehyde dehydrogenase (ALDH) 1.2.1	Mitochondria, cytosol
	Aldehyde oxidase (AO) 1.2.3.1	Cytosol
	Aldo-keto reductase (AKR)	Cytosol
	Cytochrome P450 (P450 or CYP) 1.14.13 and 1.14.14.1	ER
	Diamine oxidase (DAO) 1.4.3.6	Cytosol
	Flavin-containing monooxygenase (FMO) 1.14.13.8	ER
	Monoamine oxidase (MAO) 1.4.3.4	Mitochondria
	Prostaglandin H synthase (PGHS) 1.14.99.1	ER

(continued)

TABLE 2.2 (CONTINUED)

Pathway	Enzyme (abbreviation) EC number	Subcellular location or blood
	Xanthine oxidase/xanthine dehydrogenase (XO/XDH) 1.2.3.2/1.17.1.4	Cytosol
Hydrolysis	Carboxylesterase (CE) 3.1.1.1	ER, cytosol, lysosome
	Epoxide hydrolase (EH) 3.3.2	ER, cytosol
	β-Glucuronidase 3.2.1.31	Lysosomes, ER, blood, gut bacteria
	Arylesterases/Paraoxonases (PON) 3.1.1.2	
	Pseudocholinesterase/ butyrylcholinesterase (BuChE) 3.1.1.8	Cytosol
	Peptidase	Blood, lysosomes
Reduction	Azo- and nitro-reductase	Microflora, ER, cytosol
	Carbonyl reductase	Cytosol, blood, ER
	Disulfide reductase	Cytosol
	Quinone reductase (NADPH) 1.6.5.5	Cytosol, ER
	Sulfoxide reductase	Cytosol
	Reductive dehydrogenase	ER

EC enzyme classification; *ER* endoplasmic reticulum

TABLE 2.3. Phase II DMEs and subcellular location or blood

Enzyme (abbreviation) EC number	Subcellular location or blood
Uridine diphosphate glucuronosyltransferase (UGT) 2.4.1.17	ER
Sulfotransferase (SULT) 2.8.2	Cytosol
Glutathione *S*-transferase (GST) 2.5.1.18	Cytosol, ER
Amino acid conjugate systems	Mitochondria, ER
N-Acetyltransferase (NAT) 2.3.1.87	Cytosol
Methyltransferases	Cytosol, ER, blood

EC enzyme number; *ER* endoplasmic reticulum

In 1824, Friedrich Wohler reported the first in vivo metabolite, hippuric acid (a glycine conjugate of benzoic acid), in the urine of dogs dosed with benzoic acid.

2.3 ENZYME NOMENCLATURE

A DME is a member of a larger family of enzymes that are divided into six subclasses according to the type of reaction they catalyze (see Table 2.4). These enzymes are named by the Nomenclature Committee of the International Union of Biochemistry and Molecular Biology (http://www.chem.qmul.ac.uk/iubmb/enzyme/). In some cases, the categorization of an enzyme into one subclass or another has been challenging because of the variety of reactions catalyzed by one enzyme.

TABLE 2.4. Enzyme subclasses and the reactions they catalyze

Subclass	Type of enzyme	Reactions
EC 1	Oxidoreductases	Reduction or oxidation
EC 2	Transferases	Transfer of a functional group from one molecule to another
EC 3	Hydrolases	Hydrolysis and addition of a water molecule
EC 4	Lyases	Cleavage of various chemical bonds by means other than hydrolysis and oxidation
EC 5	Isomerases	Isomerization, racemization, and epimerization
EC 6	Ligases	Combination of two large molecules

> *The Biopharmaceutical Drug Disposition Classification System (BDDCS)* is a tool that correlates the Biopharmaceutics Classification System with the role of metabolism and transporters in the clearance of marketed drugs (see Chap. 3.5).

2.4 PHASE I: ENZYMES

2.4.1 Cytochrome P450 Enzymes (CYPs or P450s; CYP2C9. CYP2C19 and CYP3A4 EC 1.14.13; Other CYP Drug Metabolizing Enzymes EC 1.14.14.1; Non-Drug Metabolizing Enzymes are Assigned Different Numbers)

Subcellular location: Membrane-bound to the cytoplasmic side of the endoplasmice reticulum (ER; some bacterial P450s are cytosolic); mostly smooth ER.

Organ distribution: Liver, intestine, kidney, lung, and brain (see Table 2.5, for other organs see Table 2.6, for factors affecting P450 expression see Table 2.7).

Cofactor: Nicotinamide adenine dinucleotide phosphate (NADPH); electrons transferred P450 reductase and/or Cytochrome b_5.

Active site: Contains iron (Fe^{2+} or Fe^{3+}), a protoporphyrin IX ring (coordinated with iron to the four pyrrole nitrogen atoms), and a cysteine thiol coordinated to the iron as the fifth ligand.

Overall reaction: $RH + O_2 + NADPH + H^+ \rightarrow ROH + H_2O + NADP^+$, RH is the substrate and ROH is the oxidized product.

> P450 enzymes are 1.5–3% of the total microsomal protein in human livers (5% in rat livers). This is 0.3–0.6 nmol/mg of microsomal protein in humans (1 nmol/mg in rats) and 5 nmol/g in human livers (20 nmol/g in rat livers).

Steps in the P450 reaction cycle:

1. A substrate binds to Fe^{3+} (low spin state) to displace a water molecule and Fe^{3+} is changed to a high spin state.
2. Fe^{3+} accepts one electron from P450 reductase to form Fe^{2+}.
3. O_2 binds to iron.
4. A second electron transfer occurs from P450 reductase or Cytochrome b_5.
5. Formation of the reactive iron species FeO^{3+}. Other reactive iron species include FeO_2^+ and FeO_2H^{2+} (Vaz et al. 1996).
6. Hydrogen atom abstraction (or one electron abstraction) from the substrate to a radical intermediate.
7. Rebound of the hydroxyl radical to the substrate forms an oxidative metabolite (ROH).

Various examples of P450 oxidation are described in Chap. 6 (see figure 6.5) (Fig. 2.1).

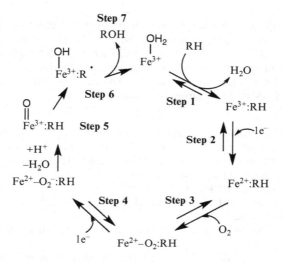

FIGURE 2.1. Cytochrome P450 (CYP) reaction cycle.

P450s exhibit an absorption maximum at 450 nm when carbon monoxide is bound to the reduced (ferrous or Fe^{2+}) form of the enzyme. A difference spectra is used for quantitating P450 by using the following equation:

$$[P450](mM) = \Delta A(at\, 450 - 490\, nm) \times 1,000/\varepsilon,$$

ε is the P450 extinction coefficient which is considered to be 91 $mM^{-1}cm^{-1}$.

Absorption at 420 nm is indicative of P450 inactivation, which is typically displacement of Fe-S from cysteine.

Reactions: See figure 6.5

Substrates and inhibitors: See Tables 5.2 and 5.4, respectively. 1-Aminobenzotriazole (ABT) is a broad inactivator of P450 isoforms. In vitro, ABT (1 mM) requires at least 15 min of preincubation in the presence of the enzyme and NADPH. In vivo, ABT inactivates P450 isoforms at po doses of 50 mg/kg in rats and 20 mg/kg in dogs and monkeys. Under these conditions, the plasma concentrations are high and are sustained for over 24 h (Balani et al. 2002). See Table 2.8 for typical characteristics of human P450 substrates.

> *P450 reductase* (containing flavoproteins in its active site) binds
> to NADPH and acts as a vehicle to transfer electrons to P450
> enzymes during the reaction cycle. The ratio of P450 reductase
> to P450 enzyme is variable, but on average could be considered
> 1:10. P450 reductase can also form reductive products, such as
> the reduction of nitroarenes to anilines.

P450 nomenclature is based on the amino acid sequence of the
enzyme. The names include a family, a subfamily, and a specific
isoform. Mouse isoforms are in lower case letters (this is not true
for any other preclinical species, see Table 2.9).

For example: CYP*3A4*

3 is the family (members of the same family share >40%
amino acid identity)

A is the subfamily (members of the same subfamily share
40–70% amino acid identity)

4 is the individual isoform in the subfamily (members of the
same individual isoform share >70% amino acid identity)

There are three main families of P450 enzymes that play a
major role in the metabolism of drugs: CYP1, CYP2, and CYP3.

TABLE 2.5. P450 isoforms and their abundance in the human liver
(Rostami-Hodjegan and Tucker 2007) and intestine (Paine 2006)

P450 isoform	Mean abundance in human liver (pmol/mg (% total))	Mean abundance in human intestine (pmol/mg (% total))	Contribution to metabolism in marketed drugs (%)[a]
CYP1A1	Not detected	5.6 (7.4%)	
CYP1A2	37 (11%)		9
CYP2A6	29 (8.6%)		
CYP2B6	7 (2.1%)		2
CYP2C8	19 (5.7%)		
CYP2C9[b]	60 (18%)	8.4 (11%)	16
CYP2C19[b]	9 (2.7%)	1.0 (1.3%)	12
CYP2D6[b]	7 (2.1%)	0.5 (0.7%)	12
CYP2E1	49 (15%)		2
CYP2J2		0.9 (1.4%)	
CYP3A4	131 (40%)	43 (57%)	46
CYP3A5		16 (21%)	

[a]The percent contribution of P450 isoforms to metabolism of the top 200
marketed drugs in 2002 (Williams et al. 2004)
[b]A polymorphic P450 isoform (see Box "Genetic polymorphisms" for
percentages)

Genetic polymorphisms are stable variations (allele variants) of the gene that encodes a DME and are observed in at least 1% of a specific population. Genetic polymorphisms are designated by the symbol * followed by a number (for example, *CYP2D6*3*. Note that genes are italicized). The numbering scheme is based on when the variant was discovered. The wild-type gene is designated as *1. Genotyping or phenotyping can be used for determining the metabolic capacity of polymorphic enzymes, which can result in changes in the pharmacokinetic properties of a drug.

The field of research that looks at the interaction between genetics and therapeutic drugs is called pharmacogenetics or pharmacogenomics.

For Caucasians, 1–3% are poor metabolizers (PMs) with respect to CYP2C9, 3–5% with respect to CYP2C19, and 5–10% with respect to CYP2D6.

TABLE 2.6. Locations of extrahepatic P450 isoforms in humans

P450 isoform	Tissue
CYP1A1	Lung, kidney, GI tract, skin, placenta
CYP1B1	Skin, kidney, prostate, mammary glands
CYP2A6	Lung, nasal membrane
CYP2B6	GI tract, lung
CYP2C	GI tract (small intestine mucosa), larynx, lung
CYP2D6	GI tract
CYP2E1	Lung, placenta
CYP2F1	Lung, placenta
CYP2J2	Heart
CYP3A	GI tract, lung, placenta, fetus, uterus, kidney

GI gastrointestinal

TABLE 2.7. Factors affecting variability in P450 expression (Rendic and Di Carlo 1997)

Factor	P450 isoforms affected
Nutrition	1A1, 1A2, 2E1, 3A4/5
Smoking	1A1, 1A2
Drugs	1A1, 1A2, 2A6, 2B6, 2C, 2D6, 3A4/5
Environment	1A1, 1A2, 2A6, 1B, 2E1, 3A4/5
Genetic polymorphism	1A, 2A6, 2C9, 2C19, 2D6, 2E1

TABLE 2.8. Typical characteristics of human P450 substrates

P450 isoform	Basic, acidic, or neutral substrates	Substrate characteristics
CYP1A2	B, N	Planar polyaromatic, one hydrogen bond donor, may contain amines or amides
CYP2A6	B, N	Small size, nonplanar, at least one aromatic ring
CYP2B6	B, N	Medium size, angular, 1–2 H-bond donors or acceptors
CYP2C8	A, N	Large size, elongated
CYP2C9	A	Medium size, 1–2 H-bond donors, lipophilic
CYP2C19	B	2–3 H-bond acceptors, moderately lipophilic
CYP2D6	B	Medium size, 5–7 Å distance between basic nitrogen and site of oxidation
CYP2E1	N	Small size, hydrophilic, relatively planar
CYP3A	B, A, N	Large size, lipophilic

Note that many exceptions exist for each of these enzymes. A list of examples of probe substrates for these enzymes is presented in table 5.2

An orthologous form of an enzyme is a similar gene product from the same evolutionary origin present in different species. Orthologous forms can be very similar, such as CYP1A2 in rat and human, or different, as in the case of CYP3A1 in rat and CYP3A4 in human. For this reason, substrate and inhibitor specificity between orthologous forms is never totally identical, and sometimes, very different.

TABLE 2.9. Major P450 isoforms in different species (Martignoni et al. 2006) (Note that mouse isoforms are written in lower case)

Isoform	Mouse	Rat	Dog	Monkey	Human
CYP1	1a1, 1a2, 1b1	1A1, 1A2, 1B1	1A1, 1A2, 1B1	1A1, 1A2, 1B1	1A1, 1A2, 1B1
CYP2A	2a4, 2a5, 2a12, 2A22	2A1, 2A2, 2A3	2A13, 2A25	2A23, 2A23	2A6, 2A7, 2A13
CYP2B	2b9, 2b10	2B1, 2B2, 2B3	2B11	2B17	2B6, 2B7
CYP2C	2c29, 2c37, 2c38, 2c39, 2c40, 2c44, 2c50, 2c54, 2c55	2C6, 2C7, 2C11[a], 2C12[a], 2C13[a], 2C22, 2C23	2C21, 2C41[b]	2C20, 2C43	2C8, 2C9, 2C18, 2C19
CYP2D	2d9, 2d10, 2d11, 2d12, 2d13, 2d22, 2d26, 2d34, 2d40	2D1, 2D2, 2D3, 2D4, 2D5, 2D18	2D15	2D17, 2D19, 2D29, 2D30	2D6, 2D7, 2D8
CYP2E	2e1	2E1	2E1	2E1	2E1
CYP3A	3a11[c], 3a13, 3a16, 3a25, 3a41, 3a44	3A1/3A23, 3A2[d], 3A9[d], 3A9, 3A18[d], 3A62	3A12, 3A26	3A8[e]	3A4, 3A5, 3A7, 3A43

[a]CYP2C11 is male-specific and is 50% of the total P450 concentration in male rats. CYP2C12 is female adult-specific. CYP2C13 is male-specific
[b]CYP2C41 is homologous with human CYP2Cs
[c]The highest activity of 3a11 is seen at 4–8 weeks of age
[d]CYP3A2 and 3A18 are male-specific and CYP3A9 is female-specific
[e]CYP3A8 is 20% of the total concentration of monkey hepatic P450 isoforms

2.4.2   Flavin-Containing Monooxygenases (FMOs; EC 1.14.13.8)

Subcellular location: Membrane-bound to the cytoplasmic side of the ER (similar to P450 enzymes).

Organ distribution: Liver, kidney, intestine, lung, brain, skin, pancreas, and secretory tissue (see Table 2.10).

Cofactor: NADPH.

Prosthetic group: Flavin adenine dinucleotide (FAD).

Reaction: See Fig. 2.2.

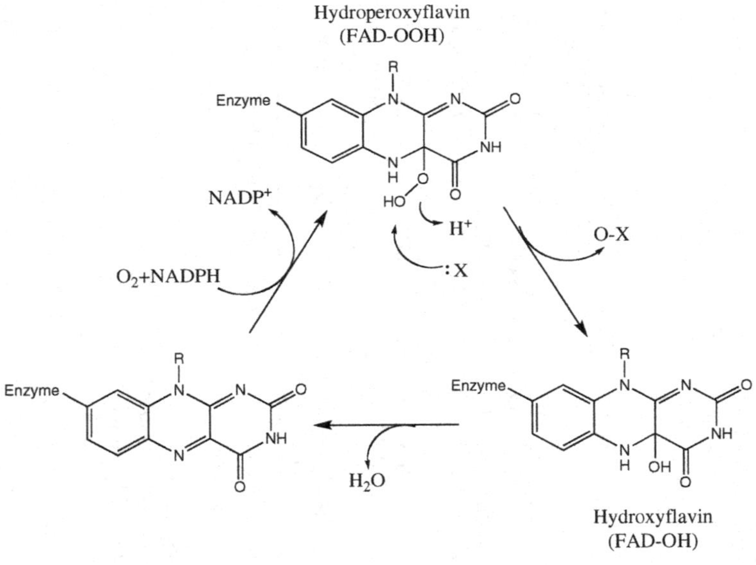

FIGURE 2.2. FMO reaction mechanism (X=N, S, P and Se). *FMO* flavin-containing monooxygenase.

Steps in the FMO reaction cycle:

1. At resting state, the enzyme is present as a hydroperoxyflavin (FAD-OOH; this form is stable). The FAD-OOH-activated intermediate can be considered a "cocked metabolism gun" because it is ready to react with a suitable substrate.
2. A nucleophilic substrate attacks the distal oxygen of FAD-OOH and results in formation of an oxygenated product and 4α-hydroxyflavin (FAD-OH).
3. The metabolite is released and FAD-OH losses water to form FAD. This step is thought to be rate-limiting.
4. FAD receives an electron from NADPH and is oxidized by O_2 to form FAD-OOH. This step formally returns FMO to its resting state.

Forms (isoforms): There are five functional forms of FMOs (FMO1 to FMO5) in humans and six nonfunctional forms. FMO3 is the major form in the human liver. FMO1 is the major form in the many other animals, but is absent in human adult livers. FMO2 is lung selective. Apparently, FMO4 and FMO5 play minor roles in drug metabolism.

- FMO1 has a shallow substrate-binding channel and, therefore, a broad substrate specificity.
- FMO3 has a deep substrate-binding channel (8–10 Å) and, therefore, a narrower specificity compared to FMO1.

Species and sex dependence in animal models. Pronounced differences in FMO expression exist in preclinical species (Janmohamed et al. 2004).

- *Female mice* have a very high expression of FMO3 and FMO5, which is considered to most closely resemble FMO expression in the adult human liver.
- *Young female rats* have higher FMO3 activity than do male rats (by five to tenfold).
- *Adult rats* have a relatively low FMO3 and FMO5 expression (unlike humans) and a constant FMO1 expression.

TABLE 2.10. FMO distribution in different species (Benedetti et al. 2006; Zhang and Cashman 2006)

Form	Mouse	Rat	Monkey	Human
FMO1	Kidney	Liver, kidney	Kidney	Kidney ≫ lung, small intestine ≫ liver
FMO2	Lung	Lung	Lung	Lung ≫ kidney > liver, small intestine
FMO3	Liver, kidney	Kidney	Liver, kidney	Liver ≫ lung > kidney ≫ small intestine

Substrates: Imipramine (FMO1), cyclobenzaprine, chlorpromazine, and nicotine (FMO3).

Inhibitors: Methimazole (also inhibits P450) and thiourea. No inhibitory antibodies are commercially available, but antibodies are available for Western Blot studies.

In humans, trimethylamine, a foul-smelling chemical derived from dietary sources such as choline and carnitine, is metabolized by FMO3 to an odorless *N*-oxide metabolite. FMO3 deficiency leads to trimethylaminuria ("fish-like odor syndrome").

FMO contribution to metabolism is usually underestimated because:

1. P450 enzymes usually generate the same metabolites.
2. Oxides (FMO or P450 metabolites) can be reduced to the parent molecule. This process is called "retro-reduction" (Cashman 2008).
3. FMOs are thermally unstable in the absence of NADPH and, therefore, can be inactivated in a preincubation step if done at 37°C (Fig. 2.3).

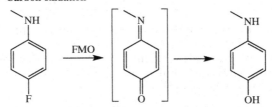

FIGURE 2.3. Examples of unusual FMO reactions: (**a**) some N-oxides are not stable, especially 3° N-oxide, and lead to Cope-type elimination, (**b**) oxime formation, and (**c**) carbon center oxidation formation of 4-hydroxylaniline from 4-fluoro-N-methylaniline (Driscoll et al. 2010). *FMO* flavin-containing monooxygenase.

Useful chemicals:
- Diethyleneamine tetra-acetic acid (DETAPAC) is used to minimize autooxidation in microsomal incubations.
- *meta*-Chloroperoxybenzoic acid (mCPBA) is useful for synthesizing oxides.
- $TiCl_3$ is used to reduce FMO *S*- and *N*-oxide metabolites.

FMOs vs. P450 enzymes

Both enzymes reside in liver microsomes (LM) and require O_2 and NADPH for their activity but:

1. *Optimum pH*: >9 for FMOs and 7.4 for P450 enzymes.
2. Most FMOs are *thermally unstable* when preincubated in the absence of NADPH. This is true for FMO1, FMO3, FMO4, and FMO5, but not FMO2.
3. *Chemical inhibitors*: Methimazole inhibits FMOs (except for FMO5) and P450 enzymes. ABT selectively inactivates P450 enzymes in a time-dependent manner, but does not inhibit FMOs.
4. Inhibitory antibodies are not available for FMOs, but are available for P450 isoforms. Antibodies to P450 can be used to distinguish between FMO-dependent and P450-dependent processes.
5. Detergents (such as Triton X-100) have little effect on FMO activity, but they inhibit P450 enzymes.
6. Recombinant enzymes can be used to distinguish between FMO- and P450-catalyzed reactions.
7. FMOs are rarely induced, but P450s are inducible.

2.4.3 Monoamine Oxidases (MAOs; EC 1.4.3.4)

Cellular location: Outer membrane of mitochondria.
Organ distribution: Expressed in most tissues.
Prosthetic group: FAD
Overall reaction:

1. RCH_2NH_2 (substrate) $+ O_2 \rightarrow RCH = NH + H_2O_2$.

2. $RCH = NH + H_2O \rightarrow RCH = O + NH_3$.

The aldehyde formed is usually oxidized further to an acid or reduced to an alcohol by other enzymes (see Table 2.11).
Isoforms: MAO-A and MAO-B possess 70% sequence identity.

Substrates: Biogenic amines; 5-Hydroxytryptamine (5-HT) (also known as serotonin) and norepinephrine (catechol-containing) are substrates of MAO-A, and 2-phenylethylamine and benzyl amine (noncatechol-containing) are substrates of MAO-B.

Inhibitors: MAO-A is inhibited by clorgyline, and MAO-B is inhibited by (*R*)-deprenyl (selegiline). Pargyline inhibits both (Fig. 2.4).

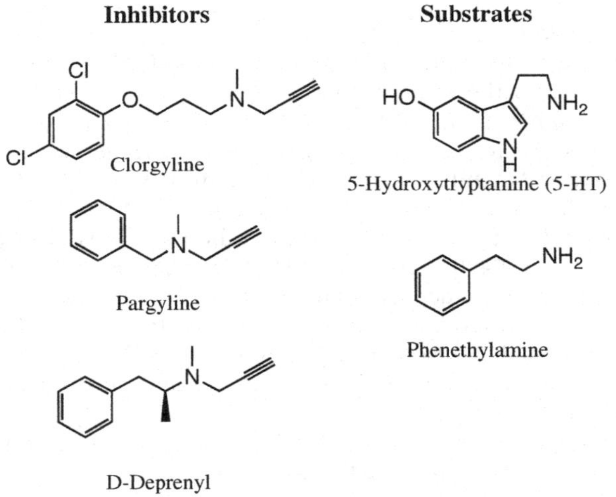

Inhibitors　　　　　　　**Substrates**

Clorgyline

5-Hydroxytryptamine (5-HT)

Pargyline

Phenethylamine

D-Deprenyl

FIGURE 2.4. Substrates and inhibitors of MAO-A and MAO-B. *MAO* monoamine oxidase.

1-Methyl-4-phenyl-1,2,3,6-tetrahydropyridine (MPTP) was inadvertently synthesized instead of the intended MPPP (a synthetic opioid) by rogue chemists. MPTP turns out to be a neurotoxin that causes permanent brain damage and Parkinsonism. The mechanism of toxicity is conversion to MPP$^+$ by MAO-B in glial cells in the brain. This metabolite kills dopamine-producing neurons in the substania nigra and, hence, induces the effects of Parkinson's disease.

2.4.4 Molybdenum Hydroxylases (AOs, XOs/XDHs)

This class includes three groups: aldehyde oxidases (AOs; EC number 1.2.3.1), xanthine oxidases (XOs; EC number 1.2.3.2), and xanthine dehydrogenases (XDHs; EC number 1.17.1.4). XDHs are the dehydrogenase forms of XOs. AOs do not have a dehydrogenase form.

Subcellular location: Cytosol.

Organ distribution: AOs are present in highest concentration in the liver, followed by the lung, kidney, and small intestine and are not present in milk or in the brain.

XOs play *a minor* role in metabolizing drugs. They are present at highest concentrations in the liver, followed by the lung, kidney, and small intestine. They are found in milk and lactating glands. Very little to no XOs are found in the brain.

Overall reaction: $RH + H_2O \rightarrow ROH + 2e^- + 2H^+$.

The source of the oxygen atom is from water and not O_2.

Active site: Contains molybdenum (Mo), Fe-S clusters, and FAD.

AO Reactions:

1. Oxidation of aliphatic and aromatic aldehydes to carboxylic acids (as the name refers, but this is a minor role it plays in humans, mostly this reaction is catalyzed by aldehyde dehydrogenases (ALDHs)).
2. Oxidation of electron-deficient sp^2-hybridized carbon next to nitrogen such as azaheterocyclic compounds and iminium ions. It was recently shown that this reaction is well predicted by determining the energetics for the intermediate formed (Torres et al. 2007).
3. Oxazole and thiazole reduction.
4. Reduction of *N*-oxides to amines (for example, brucine *N*-oxide to brucine).
5. Reduction of hydroxamic acids to amides (for example, nicotinohydroxamic acid to nicotinamide).
6. Reduction of nitro compounds to *N*-oxides (for example, conversion of benznidazole to benznidazole hydroxylamine).
7. Reduction of *S*-oxides to thiols (for example, sulindac sulfoxide to sulindac sulfide).
8. Epoxides to alkenes (e.g., benzo[a]pyrene-oxide to benzo[a] pyrene) (Fig. 2.5).

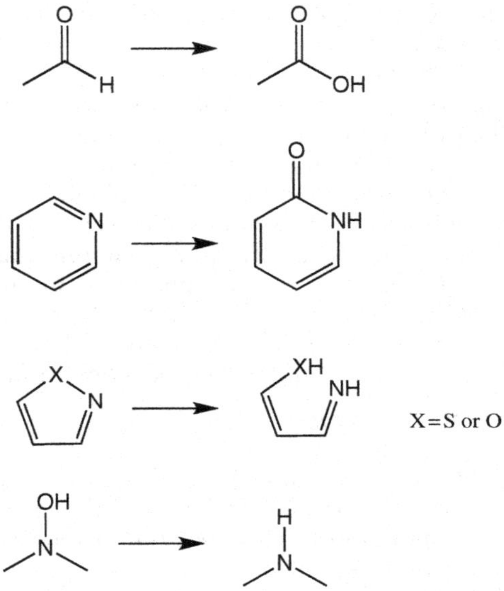

FIGURE 2.5. Various reactions catalyzed by molybdenum hydroxylases (AOs).

X = S or O

SGX523 (a cMet inhibitor) is metabolized by AOs to a 2-quinolinone metabolite that can result in crystal deposits in renal tubules, leading to renal toxicity (Diamond et al. 2010). This metabolite is formed in humans and monkeys, but not in dogs because dogs do not have any active AOs.

Isoforms: Humans have only one AO isoform, which is relatively labile. There are four AO isoforms in rodents with generally low activity, and there are no AOs in dogs. In general, AOs have broad substrate selectivity compared to XOs. AOs have more pronounced species differences than do XOs. Female rodents generally have greater AO functional activity than do male rodents. The active site of AOs in humans is the largest among the species and, therefore, accommodates more diverse substrates.

AO substrates: Phthalazine and allopurinol.

AO inhibitors: Menadione, isovanillin, raloxifene, perphenazine, thioridazine, and allopurinol (higher concentrations). Typically, substrates of AOs at higher concentrations are inhibitors.

XO substrate: Allopurinol.

XO inhibitors: Folic acid, allopurinol, and alloxanthine.

Molybdenum (Mo):
1. The human body contains about 0.07 mg of molybdenum per kg of body weight.
2. High levels of Mo can interfere with the body's uptake of copper, producing copper deficiency. Mo prevents plasma proteins from binding to copper, and also increases the amount of copper that is excreted in the urine.
3. In bacteria, molybdenum nitrogenase reduces N_2 to ammonia in a process called nitrogen fixation.

2.4.5 Alcohol Dehydrogenases (ADHs; EC 1.1.1.1)

Subcellular location: Cytosol and blood.

Organ distribution: Widely distributed, but the highest concentration is in the liver, stomach, and brain.

Cofactor: NAD^+ (Zn-containing).

Reaction: Oxidation of alcohols to aldehydes or ketones.

For alcohol oxidation, primary alcohols are preferred over secondary alcohols, most likely because of steric hindrance. In this mechanism of oxidation, a hydride (H^-) is transferred to NAD^+ to form NADH, and it is pro-R-specific.

Isoforms: High multiplicity. The major classes in humans are ADH1 to ADH5.

ADH is one of the major enzymes involved in ethanol metabolism and may have a protective effect against alcoholism.

2.4.6 Aldehyde Dehydrogenases (ALDHs; EC 1.2.1.3)

Subcellular location: Cytosol, ER, and mitochondria.

Organ distribution: Widely distributed, including in the liver, kidney, and muscle.

Cofactors: NAD^+ and $NADP^+$.

Reactions: Oxidation and reduction of aldehydes (aliphatic and aromatic), which are considered to be potentially toxic; ALDHs can also reduce quinones to hydroquinones.

Isoforms: 17 Genes are known in humans. ALDH1 is located in the cytosol and ALDH2 (oxidizes acetaldehyde) is located in the ER.

ALDH levels are elevated in tumors. ALDH3 is cytosolic tumor-specific and is involved in activation of cyclophosphamide, mitomycins, and anthracyclines to their respective toxic/active metabolites.

2.4.7   Aldo-Keto Reductases (AKRs)

Aldo-keto reductase (AKR) is a superfamily of several enzymes that catalyze reduction ketones and aldehydes to aldehydes, and reduction of quinones (Barski et al. 2008).

Subcellular location: Cytosol.

Organ distribution: Liver, mammary glands, and the brain.

Cofactors: NAD^+ and $NADP^+$.

Substrates: Typical probe substrates are menadione, glyceraldehyde, and *p*-nitrobenzaldehyde. Some endogenous substrates are steroids (androstenedione), prostaglandins, and bile acids.

2.4.8   NADPH:Quinone Reductases (NQOs; EC 1.6.5.5)

Subcellular location: Cytosol.

Organ distribution: Liver, brain, and gut. High expression in tumors. These enzymes are also targets for activation of prodrugs.

Cofactors: NADH and NADPH for NQO1; *N*-ribosyl-dihydronicotinamide (NRH) for NQO2.

Reaction: Reduction of quinones, nitroarenes, *N*-oxides, and hydroxylamines.

Inhibitors: Dicoumarol and warfarin.

TABLE 2.11. Summary of enzymes involved in the oxidation and/or reduction of alcohols, aldehydes, ketones, and carboxylic acids

Substrate	Metabolite	Enzyme (cofactor, cellular location)
Alcohol (primary)	Aldehyde	ADH (NAD^+, cytosol)
Alcohol (secondary)	Ketone	ADH (NAD^+, cytosol)
		AKR ($NAD(P)^+$, cytosol)
Alcohol (tertiary)	No reaction	
Aldehyde	Alcohol (primary)	ADH (NAD^+, cytosol)
		ALR ($NAD(P)^+$, cytosol)
		AKR ($NAD(P)^+$, cytosol)
		ALDH ($NAD(P)^+$, cytosol, ER, mitochondria)
Ketone	Alcohol (secondary)	AKR ($NAD(P)^+$, cytosol)
		Carbonyl reductase (NADPH, cytosol, ER)
		Aldehyde reductase ($NAD(P)^+$, cytosol)
Aldehyde	Carboxylic acid	ALDH
Carboxylic acid	No reduction	

ADH Alcohol dehydrogenase; AKR aldo-keto reductase; ALDH aldehyde dehydrogenase; ER endoplasmic reticulum; NAD nicotinamide adenine dinucleotide; NADPH nicotinamide adenine dinucleotide phosphate

2.4.9 Hydrolases (Hydrolytic Enzymes)

Many different enzymes possess hydrolase activity. For example, ALDH shows esterase activity. P450 enzymes also seem to have apparent hydrolase activity at times, although in the case of these enzymes the hydrolysis occurs via oxidative dealkylation by alpha oxidation next to an amide or an ester.

Overall reaction: $R_1COXR_2 + H_2O \rightarrow R_1COOH + HXR_2$
where X = O or N

Reaction: The following "catalytic triad" is important for the reaction:

- A nucleophilic residue (serine or cysteine for esterases and amidases and aspartic acid for expoxide hydrolase) that attacks the ester/thioester substrate and forms an acyl-enzyme intermediate.
- A water molecule that is coordinated by histidine and glutamic acid/aspartic acid then hydrolyzes the acyl enzyme to recycle the enzyme and liberate the hydrolysis product carboxylic acid.

Chemically, electron-deficient substrates (i.e., electron-withdrawing groups adjacent to an amide bond) result in more-labile amides susceptible to hydrolysis.

Classification based on organophosphates:

A-Esterases (also known as paraoxonases) hydrolyze organophosphates such as paraoxon. These enzymes have a free thiol (from cysteine) in their active site, which is critical for its functional activity. Inhibited by *p*-chloromercurobenzoate.

B-Esterases are inhibited by organophosphates and carbamate insecticides. Enzymes in this class include carboxylesterases and cholinesterases (acetylcholinesterase and butyrylcholinesterase).

C-Esterases include esterases that are not inhibited by organophosphates or do not hydrolyze organophosphates.

2.4.9.1 *Carboxylesterases (CEs; EC 3.1.1.1)*

Subcellular location: Mainly in the ER, cytosol, and lysosomes, lesser amounts in monocytes and macrophages.

Reaction: Hydrolysis of esters and amides.

Isoforms: hCE1 (180 kDa; optimum pH 6.5) and hCE2 (60 kDa; optimum pH 7.5–8) are the two CE isoforms in humans. hCE1 is primarily expressed in the liver, with lesser amounts in the intestine, kidney, lung, testes, and heart. hCE2 is primarily expressed in the intestine and to a lesser extent in the liver.

Substrates: Aromatic and aliphatic esters; hCE2 converts CPT-11 to SN-38.

Inhibitors: Organophosphates and 4-benzyl-piperidine-1-carboxylic acid 4-nitrophenyl ester.

Generally, rodents have a considerably greater concentration of CEs than do humans.

2.4.9.2 β-Glucuronidases (EC 3.2.1.31)

Cellular location: Lysosomes, lumen of the ER (same side as uridine diphosphate glucuronosyltransferase (UGT)), and gut bacteria.

1,4-Saccharolactone is typically used to inhibit β-glucuronidase activity in in vitro incubations. Oleson and Court argue that addition of this inhibitor does not result in improved detection of glucuronide metabolites, and in some cases, it results in UGT inhibition (Oleson and Court 2008).

2.4.9.3 Epoxide Hydrolases (EHs; EC 3.3.2)

The two classes of EH that are important for xenobiotic metabolism are microsomal EH (mEH; prefers *cis* epoxide substrates; optimum pH = 9) and soluble EH (sEH; prefers *trans* epoxide substrates; optimum pH = 7.4) (Morisseau and Hammock 2005).

Organ distribution: Both types are present in all tissues, with the highest level in the liver.

Reactions: Hydrolysis of arene and alkene epoxides to polar diols.

The hydrolysis of epoxides plays a key role in detoxification (for example, the hydrolysis of epoxide derivatives of polycyclic aromatic hydrocarbons such as benzo[a]pyrene 4,5-oxide).

mEH substrates: Benzo[a]pyrene 4,5-oxide, *cis*-stilbene oxide, and styrene oxide.

mEH inhibitors: 1,1,1-Trichloropropene-2,3-oxide, divalent heavy metals (Hg^{2+} and Zn^{2+}), and cyclopropyl oxiranes.

sEH substrate: *Trans*-stilbene oxide.

sEH inhibitors: Chalcone oxides, *trans*-3-phenylglycidols, Cd^{2+} and Cu^{2+}.

2.4.9.4 Arylesterases/Paraoxonases (PONs; EC 3.1.1.2)

Organ distribution: PON1 is synthesized in the liver and excreted to plasma and plays an important role in lipid metabolism (van Himbergen et al. 2006); PON2 is ubiquitous; PON3 is found in the liver, GI tract, kidney, lung, and brain.

Substrates: Aromatic esters phenyl acetate (PON1); *p*-nitrobutyrate (PON3 > PON1 = PON2), paraoxon (PON1), and aromatic lactones (Draganov et al. 2005).
Inhibitors: Sulfanilamide and Hg^{2+}.

2.4.9.5   *Pseudocholinesterase/Butyrylcholinesterase (BuChE; EC 3.1.1.8)*

Organ distribution: Widely distributed, high concentration in plasma.
Substrates: Basic compounds containing esters such as succinylcholine, mivacurium, procaine, and cocaine.
Inhibitors: Organophosphates and physostigmine.
BuChE was the first most widely studied gene for genetic polymorphisms.

2.5   PHASE II: ENZYMES

2.5.1   Uridine Diphosphate Glucuronosyltransferases (UGTs; EC 2.4.1.17)

Subcellular location: Luminal side of the ER.
Organ distribution: Liver, kidney, and intestine,
Cofactor: Uridine diphosphoglucuronic acid (UDPGA),
Overall reaction: See Fig. 2.6.

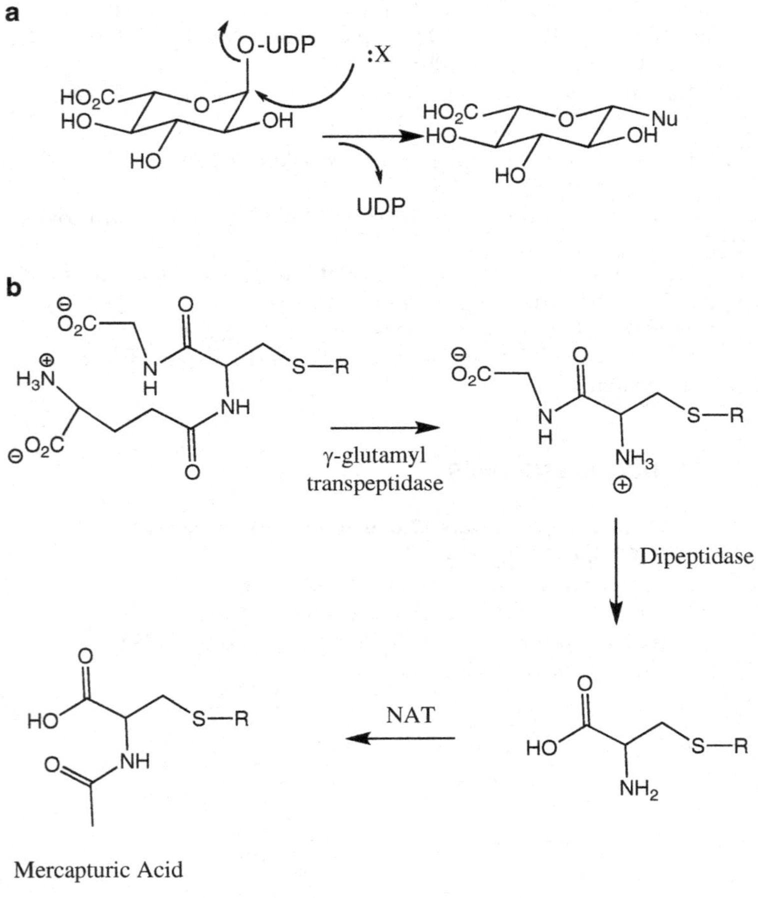

a

b

Mercapturic Acid

FIGURE 2.6. (**a**) Mechanism of formation of glucuronide conjugate. X is a nucleophilic site on the molecule such as an oxygen atom (in the case of alcohols, phenols and carboxylic acids), a nitrogen atom (in the case of amines, azaheterocyclic molecules), or a carbon atom (not common, but seen with activated carbons such as phenylbutazone). (**b**) Mechanism of breakdown of GSH conjugate to mercapturic acid (or *N*-acetyl cysteine conjugate). *UDP* uridine diphosphate; *NAT N*-acetyltransferase.

Isoforms, substrates, and inhibitors: See Table 2.12.

Typically, the glucuronides are considered to be the final metabolic products. However, other metabolic modifications have been reported, including diglucuronide conjugates. Gemifibrozil glucuronide is a potent inhibitor of CYP2C8.

A number of transporters (mostly hepatic such as MRP3 and MRP2) are known to transport glucuronide conjugates.

Sometimes, albeit rarely, glucuronides have significant pharmacological activity (i.e., morphine glucuronide).

Quantification of glucuronide metabolites:
- In vitro use of ^{14}C-UDPGA as a cofactor followed by LC separation and radioactive detection.
- Quantification of the parent drug is conducted in the presence and absence of β-glucuronidase to regenerate the aglycan parent. The difference in quantification will be the amount of glucuronide metabolite formed.

Considerations for in vitro incubations:
   Detergents (Brij 58, lubrol, and Triton X-100) or alamethicin provide access to the luminal side of the ER.

- Alamethicin is a pore-forming peptide; it should be dissolved in methanol and mixed with microsomes (20–50 µg/mg LM) on ice for 15–20 min prior to incubation.
- Detergents inhibit P450 activity, and since a number of metabolites are formed by P450 and UGT enzymes, alamethicin is used more often than detergents in in vitro incubations.

   1,4-Saccharolactone inhibits β-glucuronidase activity (see Sect. 2.3.9.2).

General Glucuronide Stability:
- Glucuronides of primary and secondary amines are acid labile.
- Acyl glucuronides are base labile.

Acyl glucuronides: A number of glucuronide conjugates of carboxylic acids (acyl glucuronides) are considered reactive electrophilic metabolites. They are capable of undergoing hydrolysis, intramolecular rearrangement, and intermolecular reactions with proteins leading to covalent drug/protein adducts. A number of acyl glucuronides lead to idiosyncratic hepatotoxicity that is considered to be immune-mediated.

TABLE 2.12. Major sites of UGT expression and the enzymes' substrates and inhibitors (Uchaipichat et al. 2006)

UGT[a]	Major sites of expression	Substrates	Inhibitors
1A1 (15%)	Liver, intestine	Bilirubin, β-estradiol, ethynylestradiol, morphine, SN-38	Atazanavir
1A3	Liver	Bile acids, cyproheptadine, alizarin, hyodeoxycholic acid	
1A4 (20%)	Liver	Imipramine, trifluoroperazine	Hecogenin
1A5	Liver, brain		
1A6	Liver, brain	Seratonin, naphthol	Amitriptyline, phenylbutazone
1A7	Stomach		Phenylbutazone, sulfinpyrizone
1A8	Intestine		
1A9	Liver, kidney	Propofol	Androsterone, phenylbutazone, sulfinpyrizone
1A10	Stomach, intestine		Amitriptyline
2B4	Liver	Xenobiotics, bile acids, hyodeoxycholic acid	
2B7 (35%)	Liver, kidney, intestine	AZT, morphine	Amitriptyline, androsterone
2B10	Liver, prostate, mammary		
2B11	Liver, kidney, prostate, adrenal		
2B15	Liver, prostate	(S)-Oxazepam	Amitriptyline, quinidine, quinine
2B17	Prostate		Amitriptyline, quinidine, quinine

AZT 3'-Azido-3'-deoxythimidine
[a]Percent of marketed drugs that are metabolized by the major UGT enzymes (Williams et al. 2004)

2.5.2 Glutathione S-Transferases (GSTs; EC 2.5.1.18)

Subcellular location: Soluble GSTs are divided into cytosolic (seven classes) and mitochondrial types. Microsomal GSTs are also known as membrane-associated proteins in eicosanoid and glutathione metabolism (MAPEG).

Cofactor: Glutathione (GSH).

Reaction: Conjugation of GSH to an electrophilic site on the substrate (see Fig. 2.6). The GST enzyme orients the substrate and the GSH in a position that increases nucleophilicity of –SH towards the substrate.

GSH conjugates are excreted in bile and its breakdown product (the N-acetylcysteine (NAC) conjugate or mercapturic acid) is excreted in urine. In the liver, the NAC conjugate is formed by dipeptidases that hydrolyze glycine and glutamate from the GSH conjugate to form a cysteine conjugate. The cysteine conjugate is transported to bile or the blood. In the kidney, N-acetyltransferase (NAT) acetylates the primary amine of cysteine to form the NAC conjugate. Also in the kidney, β-lyase can cleave the carbon–sulfur bond to form a free thiol.

Inhibitor: Ethacrynic acid.

L-Buthionine-sulfoximine (BSO) and acivicin are inhibitors of γ-glutamylcysteine synthetase and γ-glutamyltranspeptidase, respectively.

Substrates: 1-Chloro-2,4-dinitrobenzene (CDNB) and ethacrynic acid.

> GST inhibition is considered an important therapeutic area in oncology because the overexpression of this enzyme in tumors can modulate drug resistance.

2.5.3 Sulfotransferases (SULTs; EC 2.8.2)

Subcellular location: Cytosol.

Organ distribution: Liver and small intestine.

Cofactor: 3′-Phosphoadenosine-5′-phosphosulfate (PAPS).

Reaction: Transfer of sulfonate (SO_3^-) to an alcohol, amine, or hydroxylamine on the substrate.

Isoforms: SULT1A1 (expressed in the liver) conjugates small planar phenols, and SULT1A3 conjugates catecholamines.

Inhibitor: 2,6-Dichloro-4-nitrophenol (DCNP).

Substrates: Acetaminophen, minoxidil, and tamoxifen.

> SULTs typically have higher affinities for substrates than UGTs do, and therefore, SULTs are saturated at lower substrate concentrations.

2.5.4 *N*-Acetyltransferases (NATs; EC 2.3.1.87)

Subcellular location: Cytosol.

Organ distribution: In humans, NAT1 is widely distributed, and NAT2 (polymorphic) is present in the liver and small intestine.

Cofactor: Acetyl-CoA.

Reaction: Transfer of an acetyl group from acetyl-CoA to an aromatic amine or *N*-hydroxylamine.

Substrates: Aminofluorene and 4-aminobiphenyl are nonselective substrates. Sulfamethazine and isoniazide are selective NAT2 substrates.

> NAT2 is polymorphic and poor metabolizers (PMs) might have a higher incidence of adverse drug reactions and susceptibility to specific cancers.

2.5.5 Methyltransferases

Enzymes in this group are responsible for transferring a methyl group to a phenolic alcohol (such as phenol *O*-methyltransferase (POMT) and catechol-*O*-methyltransferase (COMT)), an azaheterocyclic nitrogen (such as *N*-methyltransferase), or a free thiol (such as *S*-methyltransferase).

Cofactor: *S*-adenosyl methionine (SAM).

2.5.5.1 Catechol-O-Methyltransferase (COMT; EC 2.1.1.6)

Subcellular location: Cytosol and ER.

Reaction: Transfer of a methyl group to catechols.

Inhibitors: Tolcapone and entacapone.

2.5.6 Enzymes Performing Amino Acid Conjugates

Unlike the mechanisms of other enzyme systems, amino acid conjugation involves multiple enzymes. Carboxylic acids are activated by conjugation to acetyl-CoA via acetyl-CoA synthetase (EC 6.2.1.1) present in mitochondria. This is followed by enzymatic conjugation with amino acids such as glycine, glutamate, and taurine.

References

Balani SK, Zhu T, Yang TJ et al (2002) Effective dosing regimen of 1-aminobenzotriazole for inhibition of antipyrine clearance in rats, dogs, and monkeys. Drug Metab Dispos 30:1059–1062

Barski OA, Tipparaju M, Bhatnagar A (2008) The aldo-keto reductase superfamily and its role in drug metabolism and detoxification. Drug Metab Rev 40(4):553–624

Benedetti MS, Whomsley R, Baltes E (2006) Involvement of enzymes other than CYPs in the oxidative metabolism of xenobiotics. Expert Opin Drug Metab Toxicol 2:895–921

Cashman JR (2008) Role of flavin-containing monooxygenase in drug development. Expert Opin Drug Metab Toxicol 4(12):1507–1521

Diamond S, Boer J, Maduskuie T et al (2010) Species-specific metabolism of SGX523 by aldehyde oxidase and the toxicological implications. Drug Metab Dispos 38:1277–1285

Draganov DI, Teiber JF, Speelman A et al (2005) Human paraoxonases (PON1, PON2, and PON3) are lactonases with overlapping and distinct substrate specificities. J Lipid Res 46:1239–1247

Driscoll JP, Aliagas I, Harries JJ, Halladay JS, Khatib-Shahidi S, Deese A, Segraves N, Khojasteh-Bakht SC (2010) Formation of a quinoneimine intermediate of 4-fluoro-N-methylaniline by FMO1: carbon oxidation plus defluorination Chem Res Tox 23(5):861–863

Janmohamed A, Hernandez D, Phillips IR, Shephand (2004) cell-, tissue, sex- and developmental stage-specific expression of mous flavin-containing monooxygenases (Fmos) Biochem Pharmacol 68(1)73–83

Martignoni M, Groothuis GM, de Kanter R (2006) Species differences between mouse, rat, dog, monkey and human CYP-mediated drug metabolism, inhibition and induction. Expert Opin Drug Metab Toxicol 2:875–894

Mathew N, Muthuswami Kalyanasundaram M, Balaraman K (2006) Glutathione S-transferase (GST) inhibitors. Expert Opin Ther Pat 16:431–444

Morisseau C, Hammock BD (2005) Epoxide hydrolases: mechanisms, inhibitor designs, and biological roles. Annu Rev Pharmacol Toxicol 45:311–333

Obach RS, Huynh P, Allen MC, Beedham C (2004) The human liver aldehyde oxidase: inhibition by 239 drugs. J Clin Pharmacol 44:7–19

Oleson L, Court MH (2008) Effect of the beta-glucuronidase inhibitor saccharolactone on glucuronidation by human tissue microsomes and recombinant UDP-glucuronosyltransferases. J Pharm Pharmacol 60:1175–1182

Paine MF (2006) The human intestinal cytochrome P450 "pie". Drug Metab Dispos 34:880–886

Redinbo MR, Bencharit S, Potter PM (2003) Human carboxylesterase 1: from drug metabolism to drug discovery. Biochem Soc Trans 31:620–624

Rendic S, Di Carlo FJ (1997) Human cytochrome P450 enzymes: a status report summarizing their reactions, substrates, inducers, and inhibitors. Drug Metab Rev 29(1–2):413–580

Rostami-Hodjegan A, Tucker GT (2007) Simulation and prediction of in vivo drug metabolism in human populations from in vitro data. Nat Rev Drug Discov 6(2):140–148

Torres RA, Korzekwa KR, McMasters DR et al (2007) Use of density functional calculations to predict the regioselectivity of drugs and molecules metabolized by aldehyde oxidase. J Med Chem 50:4642–4647

Uchaipichat V, Mackenzie PI, Elliot DJ et al (2006) Selectivity of substrate (trifluoperazine) and inhibitor (amitriptyline, androsterone, canrenoic acid, hecogenin, phenylbutazone, quinidine, quinine, and sulfinpyrazone) "probes" for human udp-glucuronosyltransferases. Drug Metab Dispos 34(3):449–456

van Himbergen TM, van Tits LJH, Roest M et al (2006) The story of poN1: how an organophosphate-hydrolysing enzyme is becoming a player in cardiovascular medicine. Neth J Med 64(2):34–38

Vaz ADN, Pernecky SJ, Raner GM et al (1996) Peroxo-iron and oxenoid-iron species as alternative oxygenating agents in cytochrome P450-catalyzed reactions: switching by threonine-302 to alanine mutagenesis of cytochrome P450 2B4. Proc Natl Acad Sci USA 93:4644–4648

Williams RT (1959) Detoxication mechanisms: the metabolism and detoxification of drugs, toxic substances and other organic compounds. Wiley, New York

Williams JA, Hyland R, Jones BC et al (2004) Drug–drug interactions for UDP-glucuronosyltransferase substrates: a pharmacokinetic explanation for typically observed low exposure (AUCi/AUC) ratios. Drug Metab Dispos 32:1201–1208

Zhang J, Cashman JR (2006) Quantitiative analysis of FMO gene mRNA levels in human tissues. Drug Metab Dispos 34:19–26

Zientek M, Jiang Y, Youdim K et al (2010) In vitro-in vivo correlation for intrinsic clearance for drugs metabolized by human aldehyde oxidase. Drug Metab Dispos 38:1322–1327

Additional Reading

Ortiz de Montellano PR (ed) (2004) Cytochrome P450: structure, mechanism, and biochemistry, 3rd edn. Kluwer Academic/Plenum, New York

Parkinson A, Ogilivie BW (2007) Biotransformation of xenobiotics. In: Klaassen CD (ed) Casarett & Doull's toxicology: the basic science of poisons, 7th edn. McGraw-Hill, New York

Testa B, Krämer SD (2010) The biochemistry of drug metabolism: two volume set. Wiley-VCH, Weinheim, Germany

Uetrecht JP, Trager W (2007) Drug metabolism: chemical and enzymatic aspects. Informa Healthcare, New York

Chapter 3
Oral Absorption

Abstract
The most common route of drug administration is the oral route. As such, an understanding of factors that influence oral absorption is important. For a drug to enter the systemic circulation following oral administration, the drug must first dissolve, cross the intestinal membrane, and pass through the liver. This chapter gives a general overview of oral absorption and include concepts such as first-pass metabolism.

Contents

3.1 LIST OF ABBREVIATIONS

BCS	Biopharmaceutics classification system
BDDCS	Biopharmaceutics drug disposition classification system
AUC_{PO}	AUC following an oral dose
AUC_{IV}	AUC following an intravenous dose
E_g	Extraction ratio of the gut
E_h	Extraction ratio of the liver
F	Bioavailability
F_a	Fraction of drug absorbed from the intestinal lumen
F_g	Fraction of orally administered drug escaping intestinal metabolism
F_h	Fraction of orally administered drug escaping hepatic metabolism/elimination
MAD	Maximum absorbable dose

S.C. Khojasteh et al., *Drug Metabolism and Pharmacokinetics Quick Guide*, DOI 10.1007/978-1-4419-5629-3_3,
© Springer Science+Business Media, LLC 2011

3.2 BASIC CONCEPTS

3.2.1 Bioavailability and First-Pass Metabolism

The oral bioavailability of a drug is a measure of the extent that the drug is able to enter the systemic circulation following oral administration relative to intravenous administration. Bioavailability is expressed in terms of percentage.

The calculation of bioavailability is shown in the following equation:

$$F = \frac{\text{AUC}_{\text{PO}}/\text{Dose}_{\text{PO}}}{\text{AUC}_{\text{IV}}/\text{Dose}_{\text{IV}}} \times 100 \tag{3.1}$$

For an orally administered drug to enter the systemic circulation, the drug must be absorbed through the intestinal wall, enter the portal vein, pass through the liver, and enter the systemic circulation (see Fig. 3.1).

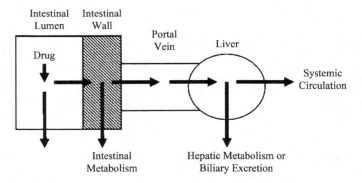

FIGURE 3.1. Movement and barriers of orally dosed drug from intestinal lumen into the systemic circulation.

As depicted in Fig. 3.1, metabolism can occur during passage through the intestinal wall and liver. Metabolism by the intestine and liver prior to reaching the systemic circulation is referred to as first-pass metabolism. Since bioavailability is a function of both absorption and first-pass metabolism of both the intestine and the liver, it can also be described by the following relationship:

$$F = F_a \times F_g \times F_h \tag{3.2}$$

where F_a is the fraction absorbed from the intestine, F_g is the fraction that escapes gut (intestinal) metabolism, and F_h is the fraction that escapes hepatic metabolism/elimination.

An alternative method of representing bioavailability in terms of extraction ratios is as follows:

$$F = F_a(1 - E_g)(1 - E_h) \qquad (3.3)$$

where E_g is the extraction ratio of the gut, and E_h is the extraction ratio of the liver.

Extraction ratios are described in Sect. 1.4.

3.2.2 Dissolution

For solid dosage forms, the process of dissolution is the first step of the oral absorption process. Dissolution is an important factor that can influence both the rate and extent of oral absorption. The Noyes–Whitney equation (see Eq. 3.4) as shown below, or modifications of it, is commonly used to describe the dissolution of compounds.

$$\text{Dissolution Rate } = \frac{dX}{dt} = \frac{AD}{h}\left(S - \frac{X_{\text{dissolved}}}{V}\right) \qquad (3.4)$$

where X is the amount of solid, $X_{\text{dissolved}}$ is the amount of dissolved drug, A is the surface area available for dissolution, D is the diffusion coefficient of the drug in the dissolution medium, h is the thickness of the diffusion layer, S is the solubility of the drug in relevant media, and V is the volume of the dissolution medium.

Based on this equation, the main properties governing drug dissolution are:

1. The solubility of the drug in gastrointestinal fluid
2. The surface area available for dissolution
3. The concentration of the drug already in solution

3.2.2.1 Solubility

Drug solubility is affected by several physicochemical and physiological factors. Solubility of a drug can be influenced by the solid state that it exists in. Crystalline forms of compound are less soluble than amorphous forms of the compound. Different crystalline forms (e.g., polymorphs, salts, solvates) may also exhibit different solubilities. Physiological factors such as pH and bile acid concentrations can also influence compound solubility. Many basic and acidic drugs often exhibit pH-dependent solubility profiles and therefore will have different solubilities in the stomach in comparison to the small intestine. Food intake can impact solubility by impacting stomach pH and also by increasing the concentration of bile acids.

3.2.2.2   Surface Area for Dissolution
The surface area for dissolution is directly correlated with particle size and shape. Dissolution rate can also be influenced by factors such as the wettability of the solid drug. Milling processes are often performed on solid drug in order to decrease particle size and improve solid drug dissolution.

3.2.3   Permeation
Following dissolution of solid drug into solution, the next step of the oral absorption process is permeation across the intestinal wall. The permeation rate is governed by the following equation:

$$\text{Permeation Rate } = \frac{\mathrm{d}X_{\text{dissolved}}}{\mathrm{d}t} = P_{\text{eff}} \times \text{SA} \times \Delta C \qquad (3.5)$$

where P_{eff} is the effective permeability of the drug to the intestinal membrane, SA is the surface area available for intestinal permeation, and ΔC is the concentration gradient across the intestinal membrane.

3.2.3.1   Permeability
Permeability is often assessed using in vitro models such as Caco-2 cells or in situ models such as the perfused rat intestine. Chapter 4 on transporters provides more details on assessment of permeability.

3.2.3.2   Surface Area
The absorption of drug occurs largely in the small intestine. The small intestine is the segment of the gastrointestinal tract that provides the largest surface area for absorption due to its numerous folds, villi, and microvilli.

3.2.3.3   Concentration Gradient
The concentration gradient that drives intestinal permeation is dependent on factors such as the dissolution rate, which controls the amount of drug in solution in the intestinal lumen, and the inherent solubility of the drug in intestinal fluid. In most cases, the plasma drug concentrations are much lower than the concentration of dissolved drug in the intestinal lumen. As a consequence, the permeation process for most drugs occurs under "sink" conditions, making dissolution the rate-limiting step for oral absorption.

3.2.4   Categories of Poor Absorption
Compounds with poor absorption can be categorized into three main categories.

3.2.4.1   Dissolution Rate-Limited Absorption

In cases of dissolution rate-limited absorption, the permeation rate is much higher than the dissolution rate. The amount of absorbed drug is dose proportional and decreases in particle size will improve oral absorption. This is typical of a BCS Class II compound (see Sect. 3.5 for definition of BCS Classes).

3.2.4.2   Permeability Rate-Limited Absorption

In cases of permeability rate-limited absorption, the dissolution rate is much higher than the permeation rate. The amount of absorbed drug is dose proportional. Any changes in particle size will not influence the amount of drug absorbed. This is typical of a BCS Class III compound.

3.2.4.3   Solubility Rate-Limited Absorption

In cases of solubility rate-limited absorption, solubility of the drug in the intestinal fluid is rate-limiting to the oral absorption process. The amount of drug absorbed is not dose proportional and any changes in particle size will not influence the amount of drug absorbed. BCS class II or IV compounds are likely to exhibit solubility rate-limited absorption.

3.2.5   Maximum Absorbable Dose

The maximum absorbable dose (MAD) provides a good conceptual tool to help anticipate oral absorption problems in the evaluation of drug candidates. It is calculated as follows:

$$\text{MAD} = k_a \times S \times V_{intestine} \times T \tag{3.6}$$

where k_a is the absorption rate constant in units of time^{-1} (i.e., min^{-1} or h^{-1}), S is the solubility of the drug in relevant media in units of concentration (i.e., mg/mL), $V_{intestine}$ is the volume of water in the small intestine in units of volume (i.e., mL), T is the small intestine transit time in units of time (i.e., min or h – this must be the same base units of time used for k_a).

If the dose is greater than the MAD, incomplete absorption may occur. The MAD serves as a conceptual tool and can be useful for the *RANK ORDERING* of compounds during the drug candidate selection process. The ability of the MAD calculation to predict the actual MAD is difficult due to challenges in determination of the relevant solubility and intestinal absorption rate constant used in the MAD calculation. Small intestine transit times required for calculation of MAD is listed for rat, dog, and human in the next section.

3.3   GASTROINTESTINAL PH AND TRANSIT TIMES

Gastrointestinal (GI) pH and transit times are useful for understanding the oral absorption of drugs (see Tables 3.1 and 3.2). As many drugs exhibit pH-dependent solubility, an understanding of the pH of the various sections of the GI tract is important. In addition, factors such as GI transit time are important for understanding the residence time of molecules in the gut and also for calculation of parameters such as MAD.

TABLE 3.1.   Gastrointestinal pH of rat, dog, and human

Species	pH (Fasted animals)		
	Stomach	Small intestine	Colon
Rat	1.93–4.15	5.89–7.10	6.23–6.70
Dog	1.80–3.00	6.20–7.3	6.45–6.75
Human	1.1–1.7	6.00–7.5	5.00–6.80

Data from GastroPlus Software (Simulations Plus, Inc., Lancaster, CA), Lui et al. (1986), and Chen et al. (2006)

TABLE 3.2.   Gastrointestinal transit times of rat, dog, and human

Species	Transit time (min)		
	Stomach	Small intestine	Entire gastrointestinal tract
Rat	15	88–109	451–844
Dog	15–96	109–110	770–844
Human	15–78	198–238	1,293–2,350

Data from Davies and Morris (1993) and GastroPlus Software (Simulations Plus, Inc., Lancaster, CA)

3.4   FOOD EFFECTS ON ORAL ABSORPTION

The effect of food on oral absorption is complex and sometimes difficult to predict. The following are a listing of some mechanisms by which food alters oral absorption:

- Food can delay gastric emptying which can result in delayed absorption.
- Food can initially raise the pH of stomach followed by a lowering of pH caused by a subsequent increase in acid secretion. This fluctuation in pH can impact the oral absorption of drugs with pH-dependent solubility profiles.

- Food can cause alterations in blood flow (i.e., splanchnic blood flow).
- Food can cause an increase in bile secretion which can help enhance the solubility of lipophilic compounds.
- Food components can chemically or physically interact with drug substance.

Food effects are greatest when the drug is administered shortly after a meal. High calorie and high fat meals are the most likely to cause a food effect.

> Pentagastrin pretreated dogs have been used to assess food effects on oral absorption. Dogs exhibit a high variability in their stomach pH. Pentagastrin treatment helps to lower basal stomach pH in dogs and in addition reduces variability in stomach pH. Lentz et al. (2007) validated this model using a set of nine compounds with known propensities for human food effect.

3.5 BIOPHARMACEUTICS CLASSIFICATION SYSTEM (BCS)

The BCS classification system is used to categorize drugs and serves to help anticipate whether drugs will have bioavailability/bioequivalence problems. BCS classifies drugs according to their solubility and permeability. A drug is considered to have high solubility if drug substance at the highest dose strength for an immediate release formulation can be dissolved in < 250 mL of water over a pH range of 1–7.5. A high permeability drug is one that has either complete intestinal absorption ($f_a > 90\%$) or exhibits rapid movement through intestinal epithelia cells in vitro. BCS classifies all drugs into four categories as shown in Table 3.3.

TABLE 3.3.
Biopharmaceutics
classification system

BCS CLASS I	BCS CLASS II
High solubility	Low solubility
High permeability	High permeability
BCS CLASS III	BCS CLASS IV
High solubility	Low solubility
Low permeability	Low permeability

BCS class I compounds (high solubility and permeability) are unlikely to show bioavailability/bioequivalence issues. Therefore, for BCS class I drugs, in vitro dissolution studies are thought to provide sufficient information to assure in vivo product performance making full in vivo bioavailability/bioequivalence studies unnecessary. BCS class II and III drugs are not eligible for biowaivers due to anticipated formulation differences in oral exposure. BCS class IV compounds, in general, are problematic with both poor solubility and permeability. The following tables (see Tables 3.4–3.7) contain lists of drugs that are categorized as BCS classes I to IV.

TABLE 3.4. BCS class I compounds (high solubility, high permeability)

Abacavir	Diazepam	Ketorolac	Phenobarbital
Acetaminophen	Diltiazem	Ketoprofen	Phenylalanine
Acyclovir	Diphenhydramine	Labetolol	Prednisolone
Amiloride	Disopyramide	Levodopa	Primaquine
Amitryptyline	Doxepin	Levofloxacin	Promazine
Antipyrine	Doxycycline	Lidocaine	Propranolol
Atropine	Enalapril	Lomefloxacin	Quinidine
Buspirone	Ephedrine	Meperidine	Rosiglitazone
Caffeine	Ergonovine	Metoprolol	Salicylic acid
Captopril	Ethambutol	Metronidazole	Theophylline
Chloroquine	Ethinyl estradiol	Midazolam	Valproic acid
Chlorpheniramine	Fluoxetine	Minocycline	Verapamil
Cyclophosphamide	Glucose	Misoprostol	Zidovudine
Desipramine	Imipramine	Nifedipine	

Adapted from Wu and Benet (2005)

TABLE 3.5. BCS class II compounds (low solubility, high permeability)

Amiodarone	Diclofenac	Itraconazole	Piroxicam
Atorvastatin	Diflunisal	Ketoconazole	Raloxifene
Azithromycin	Digoxin	Lansoprazole	Ritonavir
Carbamazepine	Erythromycin	Lovastatin	Saquinavir
Carvedilol	Flurbiprofen	Mebendazole	Sirolimus
Chlorpromazine	Glipizide	Naproxen	Spironolactone
Cisapride	Glyburide	Nelfinavir	Tacrolimus
Ciprofloxacin	Griseofulvin	Ofloxacin	Talinolol
Cyclosporine	Ibuprofen	Oxaprozin	Tamoxifen
Danazol	Indinavir	Phenazopyridine	Terfenadine
Dapsone	Indomethacin	Phenytoin	Warfarin

Adapted from Wu and Benet (2005)

TABLE 3.6. BCS class III compounds (high solubility, low permeability)

Acyclovir	Ciprofloxacin	Metformin
Amiloride	Cloxacillin	Methotrexate
Amoxicillin	Dicloxacillin	Nadolol
Atenolol	Erythromycin	Pravastatin
Atropine	Famotidine	Penicillins
Bisphosphonates	Fexofenadine	Ranitidine
Bidisomide	Folinic acid	Tetracycline
Captopril	Furosemide	Trimethoprim
Cefazolin	Ganciclovir	Valsartan
Cetirizine	Hydrochlorothiazide	Zalcitabine
Cimetidine	Lisinopril	

Adapted from Wu and Benet (2005)

TABLE 3.7. BCS class IV compounds (low solubility, low permeability)

Amphotericin B	Furosemide
Chlorthalidone	Hydrochlorothiazide
Chlorothiazide	Mebendazole
Colistin	Methotrexate
Ciprofloxacin	Neomycin

Adapted from Wu and Benet (2005)

Biopharmaceutics Drug Disposition Classification System (BDDCS)

Recently, Wu and Benet (2005) have proposed classification of drugs that differs from the BCS classification system by using the major route of elimination rather than permeability as the criteria. The categories for BDDCS is as shown below:

BDDCS CLASS I
High solubility
Extensive metabolism

BDDCS CLASS II
Low solubility
Extensive metabolism

BDDCS CLASS III
High solubility
Poor metabolism

BDDCS CLASS IV
Low solubility
Poor metabolism

Wu and Benet (2005) propose that it is easier and less ambiguous to determine BDDCS assignments based on the extent of metabolism rather than BCS assignments using permeability (i.e., extent of absorption). Also, they propose that BDDCS will facilitate predictions and expand the number of class I drugs eligible for waiver of in vivo bioequivalence studies.

References

Chen JZ, Xie M, Bao L et al (2006) Characterization of rat stomach pH following famotidine and pentagastrin pre-treatment. AAPS J 8(S2): Abstract T2301

Davies B, Morris T (1993) Physiological parameters in laboratory animals and humans. Pharm Res 10:1093–1095

Lentz KA, Quitko M, Morgan DG et al (2007) Development and validation of a preclinical food effect model. J Pharm Sci 96:459–472

Lui CY, Amidon GL, Berardi RR et al (1986) Comparison of gastrointestinal pH in dogs and humans: implications on the use of the beagle dog as a model for oral absorption in humans. J Pharm Sci 75:271–274

Rowland M, Tozer TN (2011) Clinical pharmacokinetics and pharmacodynamics: concepts and applications. Wolters Kluwer/Lippincott Williams & Wilkins, Baltimore, MD

Wu CY, Benet LZ (2005) Predicting drug disposition via application of BCS: transport/absorption/elimination interplay and development of a biopharmaceutics drug disposition classification system. Pharm Res 22:11–23

Additional Readings

Amidon GL, Lennernas H, Shah VP et al (1995) A theoretical basis for a biopharmaceutical drug classification: the correlation of in vitro drug product dissolution and in vivo bioavailability. Pharm Res 12:413–420

Ehrhardt C, Kim KJ (2008) Drug absorption studies: in situ, in vitro and in silico models. Springer Science and Business Media, LLC, New York, NY

Gu C-H, Li H, Levons J et al (2007) Predicting effect of food on extent of drug absorption based on physicochemical properties. Pharm Res 24:1118–1130

Horter D, Dressman JB (2001) Influence of physicochemical properties on dissolution of drugs in the gastrointestinal tract. Adv Drug Deliv Rev 46:75–87

Sugano K, Okazaki A, Sugimoto S et al (2007) Solubility and dissolution profile in drug discovery. Drug Metab Pharmacokinet 22:225–254

Chapter 4
Transporters

Abstract
Transporters are membrane proteins that play a role in the movement of endogenous substances and xenobiotics across cellular membranes into and out of cells. Understanding the role of transporters in drug disposition is an evolving science and over the last 15–20 years there have been great advances. As such, the transporter area is considerably less developed than what is known about common enzymes influencing drug disposition such as cytochrome P450. This chapter provides basic information on transporter families, their localization, and known substrates and inhibitors.

Contents

4.1 LIST OF ABBREVIATIONS

ABC	ATP-binding cassette transporters
ASBT	Ileal apical sodium/bile acid cotransporter
BCRP	Breast cancer resistance protein
BSEP	Bile-salt export pump
MATE	Multidrug and toxin extrusion protein
MCT	Monocarboxylic acid transporter
MRP	Multidrug resistance-associated protein
NTCP	Sodium/taurcholate cotransporting peptide

S.C. Khojasteh et al., *Drug Metabolism and Pharmacokinetics Quick Guide*, DOI 10.1007/978-1-4419-5629-3_4, © Springer Science+Business Media, LLC 2011

OAT	Organic anion transporter
OATP	Organic anion transporting polypeptides
OCT	Organic cation transporter
OCTN	Organic cation/carnitine transporter
OSTα-OSTβ	Heteromeric organic solute transporter
PEPT	Peptide transporter
P-gp	P-glycoprotein
SLC	Solute carrier transporters
URAT1	Urate transporter

4.2 BASIC CONCEPTS

4.2.1 Apical
The apical or luminal membrane is the surface of the membrane that faces the lumen.

4.2.2 Basolateral
The basolateral membrane is the surface of the membrane that forms the basal and lateral surfaces and faces away from the lumen.

4.2.3 Canalicular
The canalicular membrane is the surface of a hepatocyte membrane that faces the bile duct. The cannicular membrane is the apical membrane of a hepatocyte.

4.2.4 Sinusoidal
The sinusoidal membrane is the surface of a hepatocyte membrane facing the sinusoids and is the basolateral membrane of a hepatocyte.

4.2.5 Influx and Efflux Transporters
Influx transporters are those that move substrates into cells, whereas efflux transporters are those that pump substrates out of cells. Examples include P-glycoprotein (P-gp, MDR1, ABCB1), which is perhaps the best known efflux transporter, and organic anion transporting polypeptides (OATPs), which are examples of influx transporters.

4.2.6 Absorptive and Secretory Transporters
Absorptive transporters are transporters that transfer substrates into the systemic blood circulation. In contrast, secretory

transporters are those that transfer substrates out of the systemic blood circulation into the gut lumen, bile, or urine.

4.2.7 ABC and SLC Transporters

ATP-binding cassette (ABC) transporters and solute carrier (SLC) transporters are the two main classes that drug transporters can be categorized into. The ABC family of transporters requires ATP hydrolysis to transport substrates across membranes. Therefore, ABC transporters are primarily active transporters. Notable examples of ABC transporters include P-gp, multidrug resistance-associated protein (MRP), and breast cancer resistance protein (BCRP). In contrast to ABC transporters, SLC transporters do not have ATP-binding sites. Transport by SLC transporters use either an electrochemical potential difference in the substrate (i.e., facilitated transporters) or an ion gradient across membranes produced by primary active transporters and transport substrates against an electrochemical difference (i.e., secondary active transporters). Examples of SLC transporters include OATPs, organic anion transporters (OAT), organic cation transporters (OCT), etc. Most known drug transporters are SLC transporters.

The following are brief descriptions of select ABC transporters:

4.2.7.1   *P-Glycoprotein (P-gp, MDR1, ABCB1)*

P-gp is involved in the ATP-dependent efflux of xenobiotics from cells and is probably the best characterized of all transporters. It was discovered originally as a result of its ability to confer resistance of tumors to anticancer drugs. P-gp is found to be expressed in the intestine, kidney, liver, and the brain. It plays a role in limiting the entry of certain drugs through the blood–brain barrier. It can also play a role in intestinal absorption and in biliary and urinary excretion. Digoxin is the best characterized of the P-gp substrates.

4.2.7.2   *Breast Cancer Resistance Protein (BCRP, MXR)*

BCRP is a half ABC transporter that is expressed in the gastrointestinal tract, liver, kidney, brain, mammary tissues, testes, and the placenta. Similar to P-gp, BCRP was initially discovered due to its ability to confer resistance in cancer cell lines in vitro. BCRP plays a role in limiting oral bioavailability of certain drugs and limits entry of selected substrates through the blood–brain barrier, blood–testis barrier, and blood–placenta barrier.

The following are brief descriptions of select SLC transports:

4.2.7.3 Organic Cation Transporter (OCT) and Organic Anion Transporter (OAT)

OCTs and OATs are found in kidney and play a role in the excretion of cations and anions (xenobiotics or endogenous), respectively, into the urine. They can also be found in other tissues such as hepatocytes where they act primarily as uptake transporters and intestinal epithelia (OCT1).

4.2.7.4 Organic Anion Transporting Polypeptides (OATPs)

OATPs are involved in the sodium-independent transport of a diverse range of amphiphilic organic compounds including bile acids, thyroid hormones and steroid conjugates, and many xenobiotics. OATPs can be found in liver, intestine, kidney, and blood–brain barrier. This transporter appears to be involved in clinically relevant transporter drug–drug interactions that are of the largest magnitude.

4.2.8 Human and Rodent Nomenclature

In general, human genes and proteins are designated in capitals. Rodent genes and proteins are designated with a capital letter followed by lower case letters. For example, SLCO and OATP are the human gene and protein for OATPs. The analogous rodent gene and protein for OATPs are designated as Slco and Oatp, respectively.

4.2.9 Permeability and Efflux Ratio

Permeability is determined commonly using cell culture based models utilizing Caco-2 cells (a continuous line of heterogeneous human epithelial colorectal adenocarcinoma cells) or MDCK cells (Madin-Darby Canine Kidney Cells). In studies where the contribution of specific transporters is examined, MDCK cells over expressing a particular transporter of interest are often utilized. Examples of such MDCK cell lines include MDR1-MDCK (P-gp over expressing cells) and MRP2-MDCK II cells (MRP2 over expressing cells).

In brief, cells are grown on permeable culture inserts in transwells until they reach confluence. Compound is applied to either the apical or basolateral side of the cell monolayer depending on whether permeability is being measured from apical to basolateral or vice versa. Permeability is determined as follows:

$$P_{app} = \frac{dR}{dt} \times \frac{1}{AC_{Donor\ initial}} \qquad (4.1)$$

where

P_{app} is the apparent permeability

dR/dt is the rate of appearance of the compound in the receiver side (i.e., if compound is applied to the apical side, the basolateral side is the receiver).

A is area of the transwell insert

$C_{Donor\ initial}$ is the initial concentration on the donor side at time 0.

The efflux ratio (ER) can also be determined using an in vitro system and is defined as:

$$ER = \frac{P_{app}(BA)}{P_{app}(AB)} \qquad (4.2)$$

where P_{app} (BA) is the apparent permeability from the basolateral to apical side in an in vitro permeability assay and P_{app} (AB) is the apparent permeability from the apical to basolateral side in an in vitro permeability assay.

An ER that is >1 suggests that P_{app} (BA) > P_{app} (AB). Practically speaking, ER $\geq$3 indicate the presence of efflux. ERs can also vary with the cell system used. For example, a compound can have an ER of approximately 1 in an assay using Caco-2 cells but can have an ER >3 in a MDR1-MDCK assay. In this case, the higher ER observed in MDR1-MDCK cells may be related to the degree of over expression of MDR1 (P-gp) in the MDR1-MDCK cells.

4.3   METHODS FOR STUDYING TRANSPORTERS

4.3.1   In Vitro

ATPase Assay – A membrane assay that indirectly measures activity of the transporter of interest. The transport of substrates for ABC transporters requires ATP hydrolysis. ATP hydrolysis results in the release of inorganic phosphate which can be measured using simple colorimetric analysis.

Membrane Vesicle Assay – Inverted plasma membrane vesicles have been used to study transport. Cell lines used to prepare membrane vesicles include drug selected cells, transfected cells, and baculovirus infected insect cells. The activity of ABC transporters have been studied using this type of assay. As the membranes are inverted, influx rather than efflux is used to measure transporter activity. Using this system, detailed kinetic experiments can be performed.

Cell Lines – Polarized cell line assays where the flux of compound from apical to basolateral and basolateral to apical are commonly used to study transport and permeability. Cell lines used in this type of assay include Caco-2 (human epithelial colo-rectal adenocarcinoma cells) and MDCK (Madin-Darby canine kidney cells) cells. The expression of transporters in Caco-2 cells is comparable to that observed in the small intestine. In contrast, endogenous expression of transporters in MDCK cells is low. In addition, as MDCK cells are derived from dog, transporters expressed will be of canine origin.

Transfected Cell Lines – Transfected cell lines contain recombi-nant transporters that are either stably or transiently expressed. Transfections can be single or double and can include efflux and/or uptake transporters. Cell lines used for transfection include MDCK, LLC-PK1, HEK 293, or CHO cells. An example of a com-monly used transfected cell line is MDR1 transfected MDCK cells for the study of P-gp.

Primary Cells – Primary cells in some cases can be isolated from intact tissue and will contain the full complement of transporters present in the tissue of interest. Primary cells adapt to culture conditions quickly and transporter expression can change. An example of a primary cell assay is a brain microvessel endothelial cell assay. Properties of primary cells in culture must be under-stood prior to utilizing primary cell assays.

Hepatocyte in Sandwich Culture – A hepatocyte sandwich cul-ture assay involves culturing hepatocytes between two layers of gelled collagen. Hepatocytes in this configuration have the ability to form bile canaliculi and have a full complement of hepatic transporters on sinusoidal and canalicular membranes. Thus, this type of assay has been used to investigate biliary clearance of drugs.

4.3.2 In Vivo

Knockout models and naturally occurring transporter deficient animal models are useful tools to understand the in vivo contribu-tion of transporters. An example of the use of knockout animal studies to investigate the role of transporters on in vivo disposition is the use of $Mdr\ 1a^{-/-}$ mice to demonstrate the role of P-gp in limiting brain exposure. Despite the advantages of using in vivo systems to investigate transporters, there are limitations to these models. Alterations in the expression of other transporters and enzymes not being studied can occur in knockout animals. In addition, the level of expression of transporters can sometimes differ between animals and humans. These factors need to be considered when interpreting the results from studies using knock-out animals.

> The incorporation of the impact of transporters into physiologically based pharmacokinetic models is still in its infancy. Although the necessary theory is available, insufficient information is available on the expression of transporters in various tissues and their maximum capacity to transport various substrates in humans.

4.4 TRANSPORTER LOCALIZATION

Transporters are expressed in a variety of cells throughout the body. Figures 4.1–4.4 show transporters found in intestinal epithelial cells, hepatocytes, kidney proximal tubule cells, and brain capillary endothelial cells. Tables 4.1 and 4.2 contain information on the localization of select ABC and SLC transporters.

4.4.1 Intestine

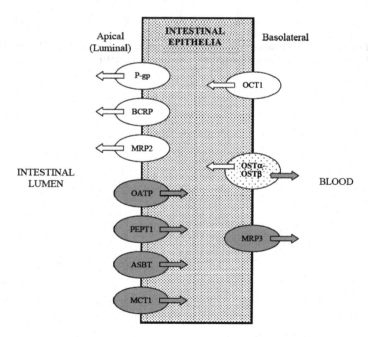

FIGURE 4.1. Localization of select transporters on intestinal epithelial cells (Adapted from The International Transporter Consortium, 2010).

4.4.2 Liver

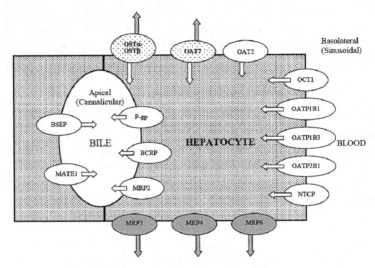

FIGURE 4.2. Localization of select transporters on hepatocytes (Adapted from The International Transporter Consortium, 2010).

4.4.3 Kidney

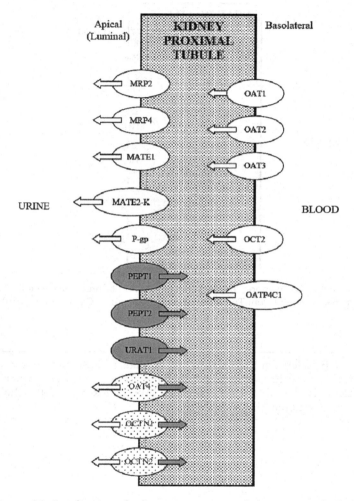

FIGURE 4.3. Localization of select transporters on kidney proximal tubule cells (Adapted from The International Transporter Consortium, 2010).

4.4.4 Blood–Brain Barrier

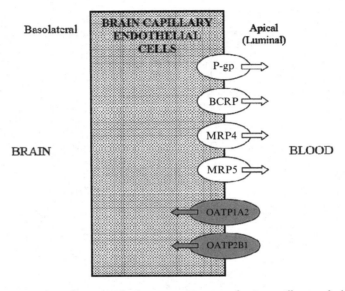

FIGURE 4.4. Localization of select transporters on brain capillary endothelial cells (Adapted from The International Transporter Consortium, 2010).

The blood–brain barrier is unique in that the basolateral membrane of brain capillary endothelial cells face the brain rather than the blood. In the intestine, liver and kidney cells, the basolateral membrane faces the blood.

TABLE 4.1. Localization of select ABC transporters

Transporter (alias)	Gene	Location (organs/cells)
MDR1 (P-gp, ABCB1)	*ABCB1*	Intestinal enterocytes, kidney proximal tubule, hepatocytes (canalicular), brain endothelia
MDR3 (ABCB4)	*ABCB4*	Hepatocytes (canalicular)
BCRP (MXR)	*ABCG2*	Intestinal enterocytes, hepatocytes (canalicular), kidney proximal tubule, brain endothelia, placenta, stem cells, mammary glands (lactating)
MRP2 (ABCC2, cMOAT)	*ABCC2*	Hepatocytes (canalicular), kidney (proximal tubule, luminal), enterocytes (luminal)
MRP3 (ABCC3)	*ABCC3*	Hepatocytes (sinusoidal), intestinal enterocytes (basolateral)
MRP4 (ABCC4)	*ABCC4*	Kidney proximal tubule (luminal), choroid plexus, hepatocytes (sinusoidal), platelets
BSEP (SPGP, cBAT, ABCB11)	*ABCB11*	Hepatocytes (canalicular)

Data from The International Transporter Consortium (2010)

TABLE 4.2. Localization of select SLC transporters

Transporter (alias)	Gene	Location (organs/cells)
OATP1A2 (OATP-A)	*SLCO1A2*	Brain capillaries endothelia, cholangiocytes, distal nephron
OATP1B1 (OATP-C, OATP2, LST-1)	*SLCO1B1*	Hepatocytes (sinusoidal)
OATP1B3 (OATP-8)	*SLCO1B3*	Hepatocytes (sinusoidal)
OATP2B1 (OATP-B)	*SLCO2B1*	Hepatocytes (sinusoidal), endothelia
OAT1	*SLC22A6*	Kidney proximal tubule, placenta
OAT3	*SLC22A8*	Kidney proximal tubule, choroid plexus, blood–brain barrier
OCT1	*SLC22A1*	Hepatocytes (sinusoidal), intestinal enterocytes
OCT2	*SLC22A2*	Kidney proximal tubule, neurons
PEPT1	*SLC15A1*	Intestinal enterocytes, kidney proximal tubule
PEPT2	*SLC15A2*	Kidney proximal tubule, choroid plexus, lung
MATE1	*SLC47A1*	Kidney proximal tubule, liver (canalicular membrane), skeletal muscle
MATE2-K	*SLC47A2*	Kidney proximal tubule

Data from The International Transporter Consortium (2010)

4.5 SUBSTRATES AND INHIBITORS

Substrates and inhibitors of transporters are, in general, less selective than substrates and inhibitors of P450 enzymes. Tables 4.3 and 4.4 summarize known substrates and inhibitors of select ABC and SLC transporters.

TABLE 4.3. Select ABC transporter substrates and inhibitors

Transporter (alias)	Gene	Substrates	Inhibitors
MDR1 (P-gp, ABCB1)	ABCB1	Digoxin, loperamide, berberine, irinotecan, doxorubicin, vinblastine, paclitaxel, fexofenadine	Cyclosporine, quinidine, tariquidar, verapamil
MDR3 (ABCB4)	ABCB4	Phosphatidylcholine, paclitaxel, digoxin, vinblastine	Verapamil, cyclosporine
BCRP (MXR)	ABCG2	Mitoxantrone, methotrexate, topotecan, imatinib, irinotecan, statins, sulfate conjugates, porphyrins	Oestrone, 17β-oestradiol, fumitremorgin C
MRP2 (ABCC2, cMOAT)	ABCC2	Glutathione and glucuronide conjugates, methotrexate, etoposide, mitoxantrone, valsartan, olmesartan, glucuronidated SN-38	Cyclosporine, delaviridine, efavirenz, emtricitabine
MRP3 (ABCC3)	ABCC3	Oestradiol-17β-glucuronide, methotrexate, fexofenadine, glucuronate conjugates	Delaviridine, efavirenz, emtricitabine
MRP4 (ABCC4)	ABCC4	Adefovir, tenofovir, cyclic AMP, dehydroepiandrosterone sulfate, methotrexate, topotecan, furosemide, cyclic GMP, bile acids plus glutathione	Celecoxib, diclofenac
BSEP (SPGP, cBAT, ABCB11)	ABCB11	Taurocholic acid, pravastatin, bile acids	Cyclosporin A, rifampicin, glibenclamide

Data from The International Transporter Consortium (2010)

TABLE 4.4. Select SLC transporter substrates and inhibitors

Transporter (alias)	Gene	Substrates	Inhibitors
OATP1A2 (OATP-A)	*SLCO1A2*	Oestrone-3-sulfate, dehydroepiandrosterone sulfate, fexofenadine, bile salts, methotrexate, bromosulphophthalein, ouabain, digoxin, levofloxacin, statins	Naringin, ritonavir, lopinavir, saquinavir, rifampicin
OATP1B1 (OATP-C, OATP2, LST-1)	*SLCO1B1*	Bromosulphophthalein, oestrone-3-sulfate, oestradiol-17β-glucuronide, statins, repaglinide, valsartan, olmesartan, bilirubin glucuronide, bilirubin, bile acids	Saquinavir, ritonavir, lopinavir, rifampicin, cyclosporine
OATP1B3 (OATP-8)	*SLCO1B3*	Bromosulphophthalein, cholecystokinin 8, statins, digoxin, fexofenadine, telmisartan glucuronide, telmisartan, valsartan, olmesartan, oestradiol-17-β-glucuronide, bile acids	Rifampicin, cyclosporine, ritonavir, lopinavir
OATP2B1 (OATP-B)	*SLCO2B1*	Oestrone-3-sulfate, bromosulphophthalein, taurocholate, statins, fexofenadine, glyburide, taurocholate	Rifampicin, cyclosporine
OAT1	*SLC22A6*	Para-aminohippurate, adefovir, cidofovir, zidovudine, lamivudine, zalcitabine, acyclovir, tenofovir, ciprofloxacin, methotrexate	Probenecid, novobiocin
OAT3	*SLC22A8*	Oestrone-3-sulfate, nonsteroidal anti-inflammatory drugs, cefaclor, ceftizoxime, furosemide, bumetanide	Probenecid, novobiocin
OCT1	*SLC22A1*	Tetraethylammonium, *N*-methylpyridinium, metformin, oxaliplatin	Quinine, quinidine, disopyramide
OCT2	*SLC22A2*		

Continued

TABLE 4.4 *Continued*

Transporter (alias)	Gene	Substrates	Inhibitors
		N-Methylpyridinium, tetraethylammonium, metformin, pindolol, procainamide, ranitidine amantadine, amiloride, oxaliplatin, varenicline	Cimetidine, pilsicainide, cetirizine, testosterone, quinidine
PEPT1	*SLC15A1*	Glycylsarcosine, cephalexin, cefadroxil, bestatin, valacyclovir, enalapril, aminolevulinic acid, captopril, dipeptides, tripeptides	Glycyl–proline
PEPT2	*SLC15A2*	Glycylsarcosine, cephalexin, cefadroxil, bestatin, valacyclovir, enalapril, aminolevulinic acid, captopril, dipeptides, tripeptides	Zofenopril, fosinopril
MATE1	*SLC47A1*	Metformin, N-methylpyridinium, tetraethylammonium	Quinidine, cimetidine, procainamide
MATE2-K	*SLC47A2*	Metformin, N-methylpyridinium, tetraethylammonium	Cimetidine, quinidine, pramipexole

Data from The International Transporter Consortium (2010)

4.6 TRANSPORTER-MEDIATED CLINICAL DRUG–DRUG INTERACTIONS

Transporter-mediated clinical drug–drug interaction occur but are less frequent than metabolic drug–drug interactions. Tables 4.5 and 4.6 summarize transporter-mediated clinical drug–drug interactions.

TABLE 4.5. ABC transporter mediated clinical drug–drug interactions

Implicated transporter	Perpetrator – victim	Pharmacokinetic impact
P-gp	Quinidine – Digoxin	Digoxin renal CL decreases 34–48%
	Ritonavir – Digoxin	Digoxin AUC increases 86%
	Dronedarone – Digoxin	Digoxin AUC increases 157% C_{max} increases 75%
	Ranolazine – Digoxin	Digoxin AUC increases 60% C_{max} increases 46%
BCRP	GF120918 – Topotecan	Topotecan AUC increases 147%

Data from The International Transporter Consortium (2010)

TABLE 4.6. SLC transporter mediated clinical drug–drug interactions

Implicated transporter	Perpetrator – victim	Pharmacokinetic impact
OATPs	Cyclosporin – Pravastatin	Pravastatin AUC increases 890% and C_{max} increase by 678%
	Cyclosporin – Rosuvastatin	Rosuvastatin AUC increase 610%
	Cyclosporin – Pitavastatin	Pitavastatin AUC increases 360% C_{max} increases 560%
	Rifampicin – Glyburide	Glyburide AUC increases 125%
	Rifampicin – Bosentan	Bosentan trough concentrations increases 500%
	Lopinavir / ritonavir – Bosentan	Bosentan Day 4 trough concentration increase 4,700%. Day 10 trough concentration increases 400%
	Lopinavir/ ritonavir – Rosuvastatin	Rosuvastatin AUC increases 107% and C_{max} increases 365%
OATs	Probenicid – Cidofovir	Cidofovir renal Cl decreases 32%
	Probenicid – Furosemide	Furosemide renal Cl decreases 66%
	Probenicid – Acyclovir	Acyclovir renal Cl decreases 32% and AUC increase 40%
OCTs	Cimetidine – Metformin	Metformin renal Cl decreases 27% and AUC increases 50%
	Cimetidine – Pindolol	Pindolol renal Cl decreases approximately 34%

Continued

TABLE 4.6 *Continued*

Implicated transporter	Perpetrator – victim	Pharmacokinetic impact
	Cimetidine – Varenicline	Varenicline AUC increases 29%
	Cimetidine – Dofetilide	Dofetilide renal Cl decreases 33%
	Cimetidine – Pilsicainide	Pilsicainide renal Cl decreases 28% and AUC increases 33%
	Cetirizine – Pilsicainide	Pilsicainide renal Cl decreases 41%

Data from The International Transporter Consortium (2010)

> With the exception of OATP-mediated drug–drug interactions, transporter-mediated drug–drug interactions are of a much less severe magnitude in comparison to metabolism-based drug–drug interactions.

References

Ehrhardt C, Kim KJ (2008) Drug absorption studies: in situ, in vitro and in silico models. Springer Science, Business Media, LLC, New York

The International Transporter Consortium (2010a) Membrane transporters in drug development. Nat Rev Drug Discov 9:215–236

You G, Morris ME (eds) (2007a) Drug transporters: molecular characterization and role in drug disposition. Wiley, Hoboken

Additional Reading

Li P, Wang G-J, Robertson TA et al (2009) Liver transporters in hepatic drug disposition: an update. Curr Drug Metab 10:482–498

Shitara Y, Horie T, Sugiyama Y (2006) Transporters as a determinant of drug clearance and tissue distribution. Eur J Pharm Sci 27:425–446

The International Transporter Consortium (2010b) Membrane transporters in drug development. Nat Rev Drug Discov 9:215–236

You G, Morris ME (eds) (2007b) Drug transporters: molecular characterization and role in drug disposition. Wiley, Hoboken

Chapter 5
Metabolism-Based Drug Interactions

Abstract
One of the main types of drug–drug interaction (DDI) is pharmacokinetic DDI, in which one drug alters the pharmacokinetics of
another drug. Here, we discuss about the two main modes of
pharmacokinetic DDIs:

1. Drug metabolizing enzyme inhibition
   (a) Reversible
   (b) Time-dependent (quasi-reversible and irreversible)
2. Enzyme induction

  The most understood and studied DDI is the metabolism-based
type. Here, we discuss inhibition, induction, and reaction phenotyping in an in vitro setting. We also discuss in vivo drug–drug
interaction predictions based on in vitro data. Useful tables of
probe substrates, selective inhibitors, and inducers are included.

Contents

5.1 LIST OF ABBREVIATIONS
AhR   Aryl hydrocarbon receptor
ALDH   Aldehyde dehydrogenase

S.C. Khojasteh et al., *Drug Metabolism and Pharmacokinetics*
Quick Guide, DOI 10.1007/978-1-4419-5629-3_5,
© Springer Science+Business Media, LLC 2011

AUC	Area under the curve
AUC$_i$	Area under the curve in the presence of an inhibitor
CAR	Constitutive androstsane receptor
C_{max}	Maximum plasma concentration
CL$_{int}$	Intrinsic clearance
DME	Drug metabolizing enzyme
E	Enzyme
f_m	Fraction metabolized by an enzyme
GST	Glutathione transferase
k	Rate constant
k_{cat}	First-order rate constant for the formation of product
k_{inact}	First-order rate constant for the formation of inactivated enzyme
K_m	Michaelis–Menten constant (i.e., [S] at $V_{max}/2$)
K_i	Inhibitory constant (competitive, Nerversible)
K_I	Concentration of inactivator (or inhibitor) at $I_{max}/2$ (irreversible)
I	Inhibitor
I_{max}	Maximal rate of inactivation
IC$_{50}$	Half maximal inhibitory concentration (i.e., [I] when $v = V_{max}/2$)
MBI	Mechanism-based inhibition
po	Oral dosing
S	Substrate
ES	Enzyme–substrate complex
P	Product
P450	Cytochrome P450
PXR	Pregnane X receptor
SULT	Sulfotransferase
TDI	Time-dependent inhibition
UGT	Uridine diphosphate glucuronosyltransferase
v	Reaction velocity
v_i	Reaction velocity in the presence of an inhibitor
V_{max}	Maximal velocity at saturated substrate concentration

5.2 BASIC CONCEPTS AND DEFINITIONS

Metabolism is one of the major routes of eliminating a drug from the body. For most drugs, only a few enzymes mediate the metabolic pathways; therefore, coadministration of drugs can potentially lead to metabolism-based DDIs. Thus, the assessment of drug interactions in the drug discovery phase, as well as in the clinic, is a necessary part of providing safe and effective drugs to the market. Many parameters are involved in this assessment, taking into consideration the role of each drug as either a victim (i.e., substrate or object) or a perpetrator (i.e., inhibitor or precipitant). For each situation, different questions need to be addressed.

For the victim (object, substrate), we need to know the contribution of various elimination pathways, the kinetic parameters for the affected pathway, the extent of plasma protein binding, and the drug concentration at the enzyme site.

For the perpetrator (inhibitor), we need to know the mechanism of inhibition, the inhibitory properties of the compound, the extent of plasma protein binding, and the inhibitor concentration at the enzyme site.

5.3 ENZYME KINETICS IN THE ABSENCE OF AN INHIBITOR

A simple enzymatic reaction is a substrate (S) binding to an enzyme (E) and forming an enzyme–substrate complex (ES). This is followed by an irreversible step to form the product (P). The rate constant k_{23} is also known as k_{cat}, which is typically the first-order irreversible step in the formation of the product (Fig. 5.1).

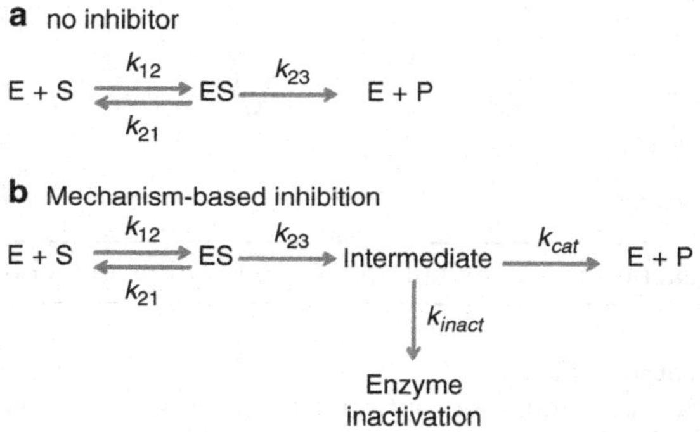

FIGURE 5.1. (**a**) Overall conversion of substrate (S) to product (P) by enzyme (E). k_{12} is a rate constant for the formation of enzyme–substrate complete (ES) and k_{21} is a rate constant for the reverse direction. k_{23} (or k_{cat}) is the rate constant for the formation of the product, which is considered irreversible. (**b**) Overall conversion of a suicide substrate (S) to inactivate the enzymes.

Assuming that the formation of the ES complex is fast relative to the rate of formation of P, the kinetics for this reaction, also known as Michaelis–Menten kinetics, is described as:

$$v = \frac{V_{max} \times [S]}{K_m + [S]} \tag{5.1}$$

v = reaction velocity.

V_{max} = maximum reaction velocity (at saturation) and equals $k_{23} \times [E]$.

$[S]$ = substrate concentration (Table 5.1).

K_m = The Michaelis–Menten constant and is defined as the substrate concentration at half V_{max} and equals $(k_{21}+k_{23})/k_{12}$ (Table 5.1).

TABLE 5.1. Linear transformations of the Michaelis–Menten equation for the diagnostics of enzyme kinetics

	Lineweaver–Burk plot	Eadie–Hofstee plot	Dixon plot
Equation	$\dfrac{1}{v} = \left(\dfrac{K_m}{V_{max}}\right)\left(\dfrac{1}{[S]}\right)$ $+ \dfrac{1}{V_{max}}$	$v = -K_m \times \left(\dfrac{v}{[S]}\right)$ $+V_{max}$	$\dfrac{[S]}{v} = \dfrac{[S]}{V_{max}} + \dfrac{K_m}{V_{max}}$
x-axis	$\dfrac{1}{[S]}$	$\dfrac{v}{[S]}$	$[S]$
y-axis	$\dfrac{1}{v}$	v	$\dfrac{[S]}{v}$
Slope	$\dfrac{K_m}{V_{max}}$	$-K_m$	$\dfrac{1}{V_{max}}$
x-intercept	$-\dfrac{1}{K_m}$	$\dfrac{V_{max}}{K_m}$	$-K_m$
y-intercept	$\dfrac{1}{V_{max}}$	V_{max}	$\dfrac{K_m}{V_{max}}$

k is a reaction rate constant and K is an equilibrium constant.

5.3.1 Practical Tips

1. Incubations should be optimized with respect to incubation time and $[E]$, preferably minimizing both.
2. Enzyme inactivation should be minimal during the incubation.
3. A pilot study should be performed using a wide range of substrate concentrations to estimate K_m. This allows for identification of enzymes with low and/or high K_m values.
4. For definitive study, a substrate concentration range from $3 \times K_m$ to $K_m/3$ could be considered. Equal intervals of the difference between the inverse values of the outer substrate concentrations usually works well. For example, if K_m is estimated at 5 µM, the concentration range would be from 15 to 1.67 µM. If six concentrations are used, then subtracting the inverse of the outer concentrations of the range from each other and dividing by six gives equal intervals of 0.0887

$((1/1.67–1/15)/6 = 0.0887)$. This translates to $[S] = 1.67$, 2.37, 3.00, 4.10, 6.43, and 15.0 µM.

5. It is important that the *consumption of the substrate* is kept at minimum during the incubation. Ideally, consumption should be <10%, but practically this limit is usually <20% especially at lower concentrations, where sensitivity for detecting the substrate may become a limitation. A number of ways to overcome this obstacle are increasing incubation time, increasing $[E]$, concentrating the samples before analysis, and monitoring metabolite formation which may be more sensitive.

6. In addition to monitoring substrate depletion, it is helpful to monitor the major metabolite(s) formed. Note that for K_m determination, no absolute quantification of substrate or metabolite(s) is necessary as the peak area ratio is enough for initial evaluation.

Sometimes the metabolite formed is a more potent inhibitor of the enzyme than the substrate, thus complicating the interpretation of the kinetics. Ideally, testing the major metabolites for inhibitory properties is helpful, but not always possible.

See Table 5.2.

TABLE 5.2. In vitro probe reactions of major P450 isoforms

P450 isoform	Reaction (K_m in µM)
CYP1A2	Phenacetin O-deethylation (47[a]); tacrine 1-hydroxylation (3–16); 7-ethoxyresorufin O-deethylation (0.2–0.5); theophylline N-demethylation (200–600); caffeine-3 N-demethylation (150–600)
CYP2A6	Coumarin 7-hydroxylation (0.84[a])
CYP2B6	Bupropion hydroxylation (82[a])
CYP2C8	Amodiaquine N-deethylation (1.9[a]); paclitaxel 6-α-hydroxylation (4–27)
CYP2C9	Tolbutamide 4-methylhydroxylation (150[a]); diclofenac 4'-hydroxylation (4.0[a])
CYP2C19	(S)-Mephenytoin 4-hydroxylation (57[a]), omeprazole 5-hydroxylation (2–6)
CYP2D6	Dextromethorphan O-demethylation (4.6[a]); bufuralol 1-hydroxylation (3–22)
CYP2E1	Chlorzoxazone 6-hydroxylation (74[a])
CYP3A	Testosterone 6β-hydroxylation (46[a]); midazolam-1'-hydroxylation (2.3[a])

[a]K_m in human liver microsomes as reported by Walsky and Obach (2004)
All other data are ranges of K_m reported since 2000 in human liver microsomes

> For determination of CYP3A inhibitory properties, two struc-
> turally diverse probe substrates are used (typically midazolam
> and testosterone), and the most potent result is used as an
> indicator of the drug's inhibitory property.

5.4   IN VITRO ENZYME INHIBITION

5.4.1   Reversible Inhibition

The four types of reversible inhibition are competitive, noncompetitive, uncompetitive, and linear mixed.

5.4.1.1   Competitive Inhibition

In competitive inhibition, the inhibitor and the substrate both bind
to the active site of the enzyme in a reversible manner.

$$v_i = \frac{V_{max} \times [S]}{(K_m \times \alpha) + [S]}, \quad \text{where } \alpha = 1 + \frac{[I]}{K_i} \tag{5.2}$$

v_i = velocity of the reaction in the presence of an inhibitor
K_i = inhibitory constant

The presence of the inhibitor does not change V_{max}, but K_m
changes by a factor of α.

IC_{50} is the inhibitor concentration that results in 50%
inhibition.

Based on Cheng and Prusoff (1973) calculations, IC_{50} for a
competitive inhibitor is as follows:

$$IC_{50} = K_i \times \left(1 + \frac{[S]}{K_m}\right) \tag{5.3}$$

If $[S] = K_m$, then $K_i = IC_{50}/2$.

5.4.1.2   Noncompetitive Inhibition

In noncompetitive inhibition, the inhibitor and the substrate bind
to the enzyme at different sites. As a result, K_m is unchanged and
V_{max} is changed by a factor of $1/\alpha$.

$$v_i = \frac{(V_{max}/\alpha) \times [S]}{K_m + [S]}, \quad \text{where } \alpha = 1 + \frac{[I]}{K_i} \tag{5.4}$$

$$IC_{50} = K_i \tag{5.5}$$

5.4.1.3 Uncompetitive Inhibition

In uncompetitive inhibition, the inhibitor binds to the ES complex. Both V_{max} and K_m are changed by a factor of $1/\alpha$.

$$v_i = \frac{(V_{max}/\alpha) \times [S]}{(K_m/\alpha) + [S]}, \quad \text{where } \alpha = 1 + \frac{[I]}{K_i} \tag{5.6}$$

$$IC_{50} = K_i \times \left(1 + \frac{K_m}{[S]}\right) \tag{5.7}$$

If $[S] = K_m$ then $K_i = IC_{50}/2$. Note that compared to the IC_{50} equation for competitive inhibition, the ratio of $[S]$ to K_m is inverted.

5.4.1.4 Linear Mixed Inhibition

In linear mixed inhibition, the inhibitor binds to both the E and ES with inhibitory constants of K_i and δK_i, respectively. V_{max} and K_m are changed by factors of $1/\beta$ and α/β, respectively (Table 5.3).

$$v_i = \frac{(V_{max}/\alpha) \times [S]}{(K_m \times \beta/\alpha) + [S]}, \quad \text{where } \alpha = 1 + \frac{[I]}{K_i} \text{ and } \beta = 1 + \frac{[I]}{\delta K_i} \tag{5.8}$$

TABLE 5.3. Effect on V_{max} and K_m under different reversible inhibition conditions

	Competitive	Noncompetitive	Uncompetitive	Linear mixed
V_{max}	$\leftrightarrow$	$\downarrow$	$\downarrow$	$\downarrow$
K_m	$\uparrow$	$\leftrightarrow$	$\downarrow$	$\uparrow$

Data analysis is conducted using a nonlinear regression program. The general rule is to input data into the different models and see which fits the data best. The simplest model that explains the data should be used. In addition, using a graphical representation is important for visual inspection of the data.

See Table 5.4.

TABLE 5.4. Preferred in vitro inhibitors of P450 isoforms

P450 isoform	Inhibitor	K_i^a (µM)	IC_{50}^b (µM)
CYP1A2	Furafylline	0.6–0.73	1.8
CYP2A6	Tranylcypromine	0.02–0.2	0.45
	Methoxsalen	0.01–0.2	
CYP2B6	2-Phenyl-2-(1-piperidinyl) propane		7.7
CYP2C8	Montelukast, Quercetin	1.1	3.1
CYP2C9	Sulfaphenazole	0.3	0.27
CYP2C19	(+)-N-3-Benzylnirvanol		0.41
CYP2D6	Quinidine	0.027–0.4	0.058
CYP2E1	Tranylcypromine		8.9
CYP3A4/5	Ketoconazole	0.0037–0.18	0.016–0.026

[a]From the preferred list of P450 inhibitors in the FDA's Draft Guidance for Industry on Drug Interaction Studies (2006)
[b]IC_{50} values in human liver microsomes reported by Walsky and Obach (2004)

5.4.2 Time-Dependent Inhibition (TDI)

Time-dependent inhibitors, as the name suggests, inhibit DME in a time-dependent manner, that is, inhibition becomes more pronounced with prolonged exposure. Mechanism-based inhibitors are a subset of time-dependent inhibitors, for which a benign (or unreactive) inhibitor is activated (i.e., metabolized) by the DME resulting in inactive enzyme (Fig. 5.1). The concern over MBI is that the total active enzyme is decreased and de novo synthesis of the enzyme is required to return to the original enzyme activity levels. Therefore, compared to competitive inhibitors, mechanism-based inhibitors impact enzyme activity even after the inhibitor is no longer present. In vitro, typically only the time-dependent, co-factor dependent, and concentration-dependent nature of the inhibition is determined and in most cases this is sufficient (see criteria for determining MBI).

There are two classes of MBI: quasi-irreversible and irreversible.

1. In quasi-irreversible inhibition, a metabolite is formed that does not easily leave the active site of the enzyme, forming a metabolic intermediate complex (MIC) and temporarily disabling the enzyme. For example, the amino groups in erythromycin and troleandomycin are oxidized by CYP3A to form nitroso groups that coordinate to the iron, thus forming an MIC.

2. In irreversible inhibition, the substrate covalently binds to the enzyme or heme alkylation.

Moieties that could lead to MBI include, but are not limited to, methylenedioxy and alkenes, acetylenes, thiophenes, furans, and alkylamines. The formation of quinone methides also could lead to MBI (see the bioactivation section in Chap. 6).

Criteria for Determining MBI

1. Inhibition of DME is time dependent.
2. Inhibition of DME is cofactor dependent (e.g., NADPH dependent in the case of P450 enzymes).
3. The presence of another substrate slows down (or removes) inactivation by competing with the MBI.
4. Inactivation is irreversible and dialysis does not retain its activity.
5. Inactivation takes places with no lag time.
6. Rate of inactivation is first order and saturable.
7. Trapping agents and reactive oxygen species (ROS) scavengers (catalase and superoxide dismutase) do not save the enzyme from inactivation. This means that MBIs do not form reactive metabolites or ROS that escape the enzyme's active site and inactivate the enzyme.

The partition ratio (r) is the ratio of k_{cat} to k_{inact}. The smaller this number the more effective the enzyme is inactivated by the MBI.

The inactivation process results in depletion of enzyme activity over time, which can be expressed by:

$$\ln\frac{E_t}{E_0} = -t \times \frac{k_{inact} \times [I]}{K_I + [I]} \tag{5.9}$$

E_t = enzyme concentration at time t
E_0 = initial enzyme concentration

5.4.2.1 Practical Tips

TDI studies are typically performed for the major CYP isoforms, or at least for CYP3A4. These assays are performed in a two-incubation process. The first incubation includes different inhibitor concentrations with enzymes in the presence or absence of cofactors. At different time points, aliquots are transferred to a new incubation that includes the probe substrate for the enzyme and, if necessary, cofactors (but no enzymes). This step is usually, but not always, a 10–20-fold dilution in order to minimize the competitive inhibition properties of the inhibitor.

One important factor to consider is the solubility of the inhibitors.

In the first step of the incubation, it is necessary to include these controls:

1. Known positive and negative controls for the enzymes
2. No inhibitor plus NADPH
3. Inhibitor but no NADPH

A nonlinear regression analysis is used to determine k_{inact} and K_I. The quality of the correlation can be checked graphically by plotting the log of the percent remaining activity (y-axis) versus incubation time (x-axis; starting from the first incubation) for each [I] (see Fig. 5.2a). The observed rate constant for inactivation (k_{obs}) is the slope of the line for each [I], which should be first order with respect to time. Then, a plot of $1/k_{obs}$ versus $1/[I]$ gives an x-intercept of $1/K_I$ and a y-intercept of $1/k_{inact}$ (see Fig. 5.2b).

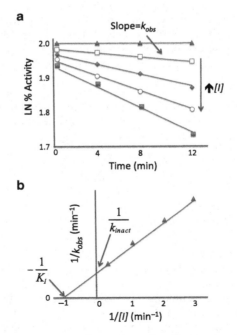

FIGURE 5.2. Concentration- and time-dependent inactivation. (**a**) natural log of percentage of activity remaining versus time. (**b**) the corresponding

$$k_{obs} = \frac{k_{inact} \times [I]}{K_I + [I]} \qquad (5.10)$$

An IC$_{50}$ *shift assay.* Determination of k_{inact} and K_I is not suitable for discovery-stage screening purposes, since it is resource intensive and time consuming. In IC$_{50}$, shift methodology is proposed where different [I] are incubated for 30 min with enzyme in the presence or absence of NADPH (Obach et al. 2007). An aliquot of each sample is then transferred to a second incubation with a probe substrate. IC$_{50}$ values are determined for each condition (+/−NADPH) by plotting percent activity of the control versus [I]. If the IC$_{50}$ is decreased (i.e., shifted) in the presence of NADPH, the compound exhibits TDI. The new IC$_{50}$ correlates well with k_{inact}/K_I (Fig. 5.3).

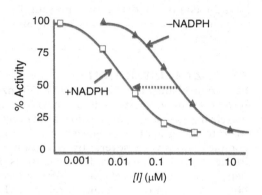

FIGURE 5.3. IC$_{50}$ shift graph: percent activity versus inhibitor concetration. Assay is run in presence and absence of NADPH after preincubation period.

Hepatocyte TDI studies may be a useful tool for considering a system with broader metabolically competent processes (Zhao et al. 2005; McGinnity et al. 2006).

See Table 5.5.

←——————————————————————————

FIGURE 5.2 *Continued.* double reciprocal plot of the rates of inactivation as a function of inhibitor concentration. k_{inact} and K_I are obtained from the reciprocal of *y*-intercept and negative *x*-intercept, respectively.

TABLE 5.5. Compounds that are mechanism-based inhibitors of P450 isoforms

P450 Isoform	Mechanism-based inhibitors
CYP1A2	Clorgyline, oltipraz, resveratrol, rofecoxib, zileuton
CYP2B6	Bergamottin, clopidogrel, ThioTEPA, ticlopidine
CYP2C8	Amiodarone, fluoxetine, gemfibrozil glucuronide, isoniazid, nortriptyline, phenelzine, verapamil
CYP2C9	Tienilic acid
CYP2C19	ThioTEPA, ticlopidine
CYP2D6	Paroxetine, serpentine
CYP3A4/5	Bergamottin, diltiazem, erythromycin, fluvoxamine, mifepristone, ritonavir, saquinavir, troleandomycin

> Furanocoumarin derivatives present in grapefruit juice are mechanism-based inhibitors of CYP3A4 in the intestine and could lead to DDI.

5.5 ENZYME INDUCTION

Several mechanisms are involved in P450 enzyme induction, which results in an increase in total enzyme concentration either by increasing the rate of expression of the enzyme or by decreasing the rate of degradation of the enzyme. These mechanisms are very complex and involve several transcriptional factors and elements in the cytosol and nucleus. Here, we focus on the critical components relevant to several major P450 isoforms: the aryl hydrocarbon receptor (AhR), pregnane X receptor (PXR), and constitutive androstane receptor (CAR). PXR and CAR regulate P450 isoforms to different extents:

PXR: CYP3A4 > CYP2B6 > CYP2C
CAR: CYP2B6 > CYP2C > CYP3A4

See Table 5.6.

TABLE 5.6. Phase I and Phase II DMEs and transporters induced by AhR, CAR, and PXR

Receptor	Phase I enzymes	Phase II enzymes	Transporters
AhR	CYP1A1, 1A2, CYP1B1, ALDH	UGT1A1, GSTA2, SULT1A1	BCRP
CAR	CYP2A, CYP2B, CYP2C, CYP3A	UGT1A1, SULT1A1	OATP2, MRP2, MRP3
PXR	CYP2B, CYP2C, CYP3A	UGT1A1, GSTA2, SULT2A1	MDR1, MRP2, OATP, OCT1

TABLE 5.7. Chemical inducers of P450 isoforms in vitro

P450 isoform	Receptor	Inducer	[Inducer] (μM)	EC_{50} (μM)	E_{max} (Fold induction)
CYP1A2	AhR	Omeprazole	25–100	0.23 ± 0.15^{a}	2.4 ± 0.9^{a}, 14–24
		β-Naphthoflavone	33–50		4–23
		3-Methylcholanthrene	1–2		6–26
CYP2B6	CAR > PXR	Phenobarbital	500–1,000	58 ± 96^{a}	7.6 ± 1.8^{a}, 5–10
CYP3A4/5	PXR > CAR	Rifampin	10–25	0.85 ± 0.75^{a}	12 ± 3^{a}, 4–31

[a]Kato et al. 2005 Drug Metabolism Pharmacokinet 20:236–243
All the other data from FDA's Draft Guidance for Industry on Drug Interaction Studies (2006)

Practical Tips

1. The compound must be compared to an "appropriate" positive control (see Table 5.7).
2. A compound is considered an inducer if it produces a response $\geq$2-fold over the base line.
3. Clinically relevant drug concentrations (at least three concentrations) or full dose response profiles should be used with at least one order of magnitude higher than C_{max}.
4. Typically, CYP activity is used as a marker. Correlation between enzyme induction and increased mRNA levels have been established and it is more sensitive than activity determination. Also, when assessing a time-dependent inhibitor, CYP activity could be masked by the induction but mRNA levels are more indicative.
5. At least three different human hepatocytes individual donars should be tested.
6. Cryopreserved hepatocytes as well as freshly plated hepatocytes could be used.
7. Other cell-based assays used for this purpose are (but not limited) HepG2 cell transfected with CYP3A4 and luciferase, Fa2N-4, and HepaRG.
8. It is important to assess cytotoxicity at every concentration of the test compound.

5.6 REACTION PHENOTYPING

Reaction phenotyping is used to identify the major enzyme(s) involved in the metabolism of a drug. This information allows for determination of f_m, which leads to further understanding of possible interpatient pharmacokinetic variability and/or the potential for drug–drug interactions when the drug is being considered as a victim. Ultimately, results from human radiolabelled mass balance studies are needed to complement in vitro reaction phenotyping studies to allow for a true determination of f_m for each metabolic pathway.

Note: When to perform reaction phenotyping:

1. Metabolism has to be a major route of elimination of the drug from the body. This can be inferred from preclinical studies, based on when unchanged drug in the excreta is determined to be minimal.
2. The major metabolites in vivo are known. This is because the right enzyme system is studied in vitro. For example, if the

primary route of metabolism is via glucuronidation of the unchanged drug, P450 reaction phenotyping may not be very useful.

5.6.1 Practical Tips

1. Studies are usually performed at 1 µM, but once the enzyme(s) involved has been identified, performing the incubations at multiple concentrations is useful for determining the role of enzymes with different K_m and V_{max}.
2. In vitro and in vivo metabolite identification are necessary parts of reaction phenotyping to ensure that the right enzymes are being examined.
3. Typically performed monitoring drug disappearance from the incubation. When CL_{int} is low, it is important that the metabolites are monitored.

> *The rate of metabolism* is how fast a compound is metabolized.
> *The extent of metabolism* is the contribution of metabolism by a certain pathway to the overall CL.
> So a compound can have a slow rate of metabolism, yet the extent of metabolism can be high.

5.6.1.1 P450-Based Reaction Phenotyping

Even though reaction phenotyping can be conducted for various enzymes, these studies are most often used for P450 enzymes. For other enzyme systems, the recombinant enzymes and inhibitors that can be used are listed in Chap. 2.

5.6.1.2 Practical Tips

1. 1-Aminobenzotriazole (ABT; 1 mM) is preincubated for 15–30 min with liver microsomes (0.5–2 mg/mL) or hepatocytes (0.5–1 × 10^6 cells/mL) to inactivate P450 activity. The metabolites formed are not P450 dependent.
2. When using recombinant human P450 isoforms, equal concentrations of each isoform are usually used. Based on the quantities of P450 isoforms in an average liver, the contribution of the isoforms can be recalculated (see Table 2.5).
3. Both chemical and antibody selective inhibitors can be used with liver microsomes. Typically, chemical inhibitors are used, perhaps because of their low cost.
4. Correlation analysis studies in multiple liver microsomes with characterized P450 isoform activities are not often conducted.

This is for several reasons, including complications that can occur when multiple enzymes are involved.

In general, any metabolic pathway that contributes to $f_m \leq 50\%$ of the CL_{int} of a drug does not contribute to drug interactions with AUC fold change <2 (see Table 5.8).

For polymorphic enzymes that are a major component of the CL_{int}, a clinical study comparing poor and extensive metabolizers is required to determine variability in the extent of drug interactions with an inhibitor in different populations.

5.7 PREDICTING IN VIVO DRUG INTERACTIONS

In vitro data are used to predict the magnitude of in vivo metabolism-based DDIs in the clinic. No animal in vivo model is entirely suitable for assessing DDI risk in humans due to many factors, including differences in enzymes and disposition of the compound. Potential drug interactions are assessed by taking into consideration the drug as a victim or a perpetrator.

There are limitations to quantitatively predict in vivo DDIs due to factors such as the involvement of transporters and uncertainty in the concentrations of the substrate and inhibitor at the active site of the enzyme. However, in addition to DDI prediction, it is possible to use in vitro data to rank order the isoforms according to inhibitory potency. This allows for clinical DDI studies to be performed first on the most potently inhibited enzyme, the outcome of which then informs further studies. This approach may prevent potentially unnecessary clinical drug interaction studies if no clinical inhibition is observed in the first study.

In the following sections, we discuss the static prediction models used for competitive and mechanism-based inhibitors and inductors. Dynamic models are currently available using Simcyp® software.

5.7.1 In Vivo Prediction of DDIs for Competitive Inhibitors

The relationship between the magnitude of DDI and increase in drug exposure in the liver follows the following expression:

$$\text{Liver DDI magnitude} = \frac{\text{AUC}_i}{\text{AUC}} = \frac{1}{\left(\frac{f_m}{\alpha}\right) + (1 - f_m)} = A(\text{liver})$$

$$\text{where } \alpha = 1 + \frac{[I]}{K_i}. \tag{5.11}$$

f_m = fraction of the drug metabolized by the inhibited enzyme
$[I]$ = in vivo inhibitor concentration

The value of $[I]$ in this equation has great uncertainty. Theoretically, this parameter is the concentration of the inhibitor at the site of the enzyme in the liver. The unbound hepatic inlet C_{max} is the most predictive concentration ($C_{hep,inlet,u}$; Obach et al. 2006) and is derived in the following way (Kanamitsu et al. 2000):

$$C_{hep,inlet,u} = C_{max,u} + \frac{f_u \times D \times k_a \times F_a}{Q_h} \qquad (5.12)$$

D = dose
k_a = absorption rate constant
F_a = fraction of unchanged inhibitor passing through the intestine
Q_h = hepatic blood flow (21 mL/min/kg for humans).

TABLE 5.8. Impact of f_m on change in DDI magnitude ($AUC_{po,i}/AUC_{po}$) at $[I]/K_i = 1, 10, 100$

f_m	Fold change in magnitude of DDI ($AUC_{po,i}/AUC_{po}$)		
	$[I]/K_i = 1$	$[I]/K_i = 10$	$[I]/K_i = 100$
0.2	1.11	1.22	1.25
0.4	1.25	1.57	1.66
0.6	1.43	2.2	2.46
0.8	1.67	3.67	4.81
1	2.00	11.0	101

For compounds metabolized by intestinal P450 isoforms (mainly CYP3A), metabolism-based DDI in the intestine is also considered.

$$\text{Intestinal DDI magnitude} = \frac{1}{\left(\frac{1-F_G}{\alpha}\right) + F_G} = A(\text{intestine})$$

$$\text{where } \alpha = 1 + \frac{[I]}{K_i} \qquad (5.13)$$

F_G = intestinal bioavailability of the substrate

Total DDI magnitude for a competitive inhibitor = A(intestine) × A(liver).

In the FDA Draft Guidance for Industry on Drug Interaction Studies (2006), the following equation is used for competitive inhibition:

$$\text{Magnitude of DDI} = 1 + \frac{[I]}{K_i} \tag{5.14}$$

where $f_m = 1$ in the above equation and suggest a conservative $[I]$ as a mean steady-state C_{max} (bound plus unbound) following administration of the highest proposed clinical dose (Table 5.9).

TABLE 5.9. Prediction of clinical DDIs based on the FDA draft guidance for Industry on Drug Interaction Studies (2006)

$[I]/K_i$	Potential of clinical DDIs
<0.1	Remote
0.1–1	Possible
>1	Likely

5.7.2 In Vivo Prediction of DDIs for Mechanism-Based Inhibitors

Mechanism-based inhibitors inactivate DMEs; therefore, the enzyme has to be resynthesized in order to recover to its original activity. In the liver and intestine, there is a constant rate of enzyme synthesis (k_{syn}) and enzyme degradation (k_{deg}). In the absence of a mechanism-based inhibitor and at steady-state, $k_{syn} = k_{deg}$. In the presence of a mechanism-based inhibitor, the total enzyme degradation rate is increased from its natural rate (k_{deg}) by k_{inact}. The net result is a transient decrease in the total active enzyme concentration. The time it takes for the enzyme to recover to its original enzyme level depends on the de novo synthesis rate, which is different for each isoform and depends on the organ (liver versus intestine).

DDI magnitude (AUC$_i$/AUC) in the presence of an MBI is determined using the following equation (based on Mayhew et al. 2000):

$$\text{DDI magnitude (MBI)} = \frac{1}{\left(\dfrac{f_m}{\gamma}\right) + (1 - f_m)} \tag{5.15}$$

$$\gamma = 1 + \frac{k_{inact} \times [I]}{k_{deg} \times ([I] + K_I)}$$

f_m = fraction of the substrate metabolized by the enzyme
$[I]$ = inhibitor concentration
K_I = concentration of inactivator (or inhibitor) at $I_{max}/2$
k_{inact} = first-order rate constant for the formation of inactivated enzyme
k_{deg} = rate constant for the natural rate of degradation of the enzyme (Table 5.10).

TABLE 5.10. Half-life and k_{deg} of human hepatic CYP isoforms and intestinal CYP3A4

P450 isoform	Half-life (h)	k_{deg} (h^{-1})
CYP1A2	36–105	0.0066–0.01926
CYP2B6	32	0.0217
CYP2C8	23 (8–41)	0.0301 (0.0169–0.0864)
CYP2C9	104	0.00666
CYP2C19	26 (7–50)	0.0266 (0.0139–0.099)
CYP2D6	51, 70	0.0136, 0.0099
CYP3A4	44–140	0.00495–0.0158, 0.0077
CYP3A5	36 (15–70)	0.0193 (0.0099–0.0462)
CYP3A4 (intestinal)	12–33	0.021–0.0578

All the data based on Yang et al. (2008) except for CYP3A4 of $k_{deg} = 0.0077$ h^{-1} from Wang 2010

5.7.3 In Vivo Prediction of DDIs for Inducers

The magnitude of a DDI involving an inductor is based on the equation below:

$$\text{DDI magnitude (induction)} = \frac{1}{\left(\dfrac{f_m}{\varepsilon}\right) + (1 - f_m)} \tag{5.16}$$

$$\varepsilon = 1 + \frac{\text{sf} \times E_{max} \times [I]}{[I] + EC_{50}} \tag{5.17}$$

sf = a scaling factor that is empirically derived
E_{max} = maximum induction response
$[I]$ = concentration of inducer
EC_{50} = concentration of inductor that results in half E_{max}

If a compound is a competitive inhibitor, a mechanism-based inhibitor, and an inducer, then the magnitude of DDI is expressed below:

$$\text{DDI magnitude} = \frac{1}{\left(\dfrac{f_m}{\alpha \times \beta \times \varepsilon}\right) + (1 - f_m)} \tag{5.18}$$

Ritonavir is a competitive inhibitor, mechanism-based inhibitor, and inducer of CYP3A4.

See Tables 5.11 and 5.12.

TABLE 5.11. Comparison of different drug interactions

Mechanism	Reversible inhibition	Mechanism-based inhibition	Induction
Onset	Immediate	Depends on the rate of enzyme inactivation	Slow (days)
Is prior exposure to inhibitor needed?	Not needed	Needed	Needed
AUC (for victim)	Increased	Increased	Decreased

FDA Decision Tree Base on In Vitro Studies

Substrate:

- If not a substrate → stop and label accordingly
- If a substrate → conduct in vivo studies with potent inhibitor (s)/inducer(s) → adjust dose and/or label accordingly

Inhibitor/inducer:

- If not an inhibitor/inducer → stop and label accordingly
- If an inhibitor/inducer → conduct in vivo studies, adjust dose and/or label accordingly

TABLE 5.12. Human oral in vivo substrates (f_m for that particular enzyme), inhibitors and inducers for P450 isoforms

P450 isoform	Substrate (f_m)	Inhibitor	Inducer
CYP1A2	Theophylline, caffeine	Fluvoxamine	Smoking
CYP2B6	Efavirenz	Clopidogrel	Rifampin
CYP2C8	Repaglinide (0.49^a), rosiglitazone (0.5^a)	Gemfibrozil	Rifampin
CYP2C9	Warfarin (0.85^b), tolbutamide (0.85^b)	Fluconazole, amiodarone	Rifampin
CYP2C19	Omeprazole, esomprazole, lansoprazole, pantoprazole	Omeprazole, fluvoxamine, moclobemide	Rifampin
CYP2D6	Desipramine (0.97^c, 0.877^d), dextromethorphan, atomoxetine	Paroxetine, quinidine, fluoxetine	Not induced
CYP2E1	Chlorzoxazone	Disulfiram	Ethanol
CYP3A	Midazolam (0.94^e), buspirone (0.99^f), felodipine (0.81^e), lovastatin (0.9^g, 0.99^a), eletriptan, sildenafil (0.9^h), simvastatin (0.99^a), triazolam (0.92^e)	Atazanavir, clarithromycin, indinavir, itraconazole, ketoconazole, nefazodone, nelfinavir, ritonavir, saquinavir, telithromycin	Rifampin, carbamazepine

Source: http://www.fda.gov/Drugs/DevelopmentApprovalProcess/
DevelopmentResources/DrugInteractionsLabeling/ucm081177.htm#inVivo
visited last August 1, 2010
[a]Hinton LK et al. (2008) Pharm Res 25:1063–74
[b]Miners JO, Birkett DJ (1998) Br J Clin Pharmacol 45:525–538
[c]Rowland YK et al. (2004) Drug Metab Dispos 32:1522
[d]Ito et al. (2005) Drug Metab Dispos 33:837–844
[e]Brown HS et al. (2005) Br J Clin Pharmacol 60(5):508–18
[f]Galetin A et al. (2006) Drug Metab Dispos 34:166–75
[g]Shitara Y, Sugiyama Y (2006) Pharmacol Ther 112:71–105
[h]Houston, Galetin (2008) Current Drug Metab (2008) 9:940–951

References

Cheng YC, Prusoff WH (1973) Relationship between the inhibition constant (&) and the concentration of inhibitor which causes 50 per cent inhibition (iso) of an enzymatic reaction. Biochem Pharmacol 22:3099–3108

FDA Draft Guidance for Industry on Drug Interaction Studies–Study design, data analysis, and implication for dosing and labeling. September 2006

Kanamitsu SI, Ito K, Sugiyama Y (2000) Quantitive prediction of in vivo drug–drug interactions from in vitro data based on physiological pharmacokinetics:use of maximum unbound concentration of inhibitor at the inlet to the liver. Pharm Res 17:336–343

Mayhew BS, Jones DR, Hall SD (2000) An in vitro model for predicting in vivo inhibition of cytochrome P450 metabolic intermediate complex formation. Drug Metab Dispos 28:1031–1037

McGinnity DF, Berry AJ, Kenny JR, Grime K, Riley RJ (2006) Evaluation of time-dependent cytochrome P450 inhibition using cultured human hepatocytes. Drug Metab Dispos 34:1291–1300

Obach RS, Walsky RL, Venkatakrishnan K, Gaman EA, Houston JB, Tremaine LM (2006) The utility of in vitro cytochrome P450 inhibition data in the prediction of drug–drug interactions. J Pharmacol Exp Ther 316:336–348

Obach RS, Walsky RL, Venkatakrishnan K (2007) Mechanism-based inactivation of human cytochrome P450 enzymes and the prediction of drug–drug interactions. Drug Metab Dispos 35:246–255

Walsky RL, Obach RS (2004) Validated assays for human cytochrome P450 activities. Drug Metab Dispos 32:647–660

Wang Y-H (2010) Confidence assessment of the Simcyp time-based approach and a static mathematical model in predicting clinical drug–drug interactions for mechanism-based CYP3A inhibitors. Drug Metab Dispos 38:1094–1104

Yang J, Liao M, Shou M, Jamei M, Yeo KR, Tucker GT, Rostami-Hodjegan A (2008) Cytochrome P450 turnover: regulation of synthesis and degradation, methods for determining rates, and implications for the prediction of drug interactions. Curr Drug Metab 9:384–393

Zhao P, Kunze KL, Lee CA (2005) Evaluation of time-dependent inactivation of CYP3A in cryopreserved human hepatocytes. Drug Metab Dispos 33:853–861

Further Readings

Cozza KL, Armstrong SC, Oesterheld JR (2003) Concise Guide to Drug Interaction Principles for Medical Practice: Cytochrome P450s, Ugts, P-Glycoproteins, 2nd edn. American Psychiatric Publishing, Washington, DC

Kenny JR, McGinnity DF, Grime K, Riley RJ (2008) Utilizing of in vitro Cytochrome P450 inhibition data for projection clinical drug-drug interactions. Wiley, New York

Levy RH, Thummel KE, Trager WF (2000) Metabolic Drug Interactions. Lippincott Williams & Wilkins, Philadelphia

Rodrigues AD (2008) Drug-Drug Interactions, 2nd edn. Marcel Dekker, New York

Segel IH (1993) Enzyme Kinetics: Behavior and Analysis of Rapid Equilibrium and Steady-State Enzyme Systems. Wiley, New York

Walsky RL, Boldt SE (2008) In vitro cytochrome P450 inhibition and induction. Curr Drug Metab 9:1–12

Chapter 6
Biotransformation and Bioactivation

Abstract

The terms metabolism and biotransformation are used inter-
changeably in this book. As described in Chap. 2, metabolism is a
major route of elimination of drugs from the body and, in general,
results in the formation of metabolites that are more polar than the
parent drug. Here, we discuss techniques for detection of metabolites, bioactivation and its ramifications, metabolites in drug safety
studies, and what transformation happens to different common
moieties in drugs as a result of metabolism.

Contents

6.1 LIST OF ABBREVIATIONS

DME	Drug metabolizing enzyme
DBE	Double bond equivalent
GSH	Glutathione
GST	Glutathione S-transferase
H/D	Hydrogen/deuterium
HAT	Hydrogen atom transfer
ICH	International Conference on Harmonisation
LC	Liquid chromatography
MIST	Metabolites in safety testing
MS	Mass spectroscopy

S.C. Khojasteh et al., *Drug Metabolism and Pharmacokinetics*
Quick Guide, DOI 10.1007/978-1-4419-5629-3_6,
© Springer Science+Business Media, LLC 2011

NADPH   Nicotinamide adenine dinucleotide phosphate
NAPQI N-acetyl-p-quinoneimine
NMR   Nuclear magnetic resonance
P450   Cytochrome P450
PD   Pharmacodynamic
PK   Pharmacokinetic
SET   Single electron transfer
UV   Ultraviolet

6.2 OVERVIEW OF BIOTRANSFORMATION

Various drug metabolizing enzymes (DMEs) lead to formation of metabolites. Metabolites can be:

- Inactive towards the target, as is seen in most cases
- Active, either on-target, such as fluoxetine→norfluoxetine, or off-target
- Toxic, such as acetaminophen→N-acetyl-p-quinoneimine (NAPQI) at high doses
- The ultimate drug, such as the conversion of prodrug (e.g., 5-fluorouracil)→drug (e.g., fluorouridine triphosphate and fluorodeoxyuridine triphosphate).

6.3 METABOLITE DETECTION AND IDENTIFICATION

Various analytical tools can detect and characterize metabolites (Liu and Hop 2005). Typically, LC-MS/MS and UV methods are used and, to a lesser extent, NMR. In Chap. 8, we discuss the various MS detection methods for accomplishing this task.

6.3.1 MS Data (Full Scan)

In full scan, changes in mass to a molecular ion are used to determine changes to the parent compound (see Table 6.1). This information, plus knowing the source of the metabolite, allows for determination of overall changes to the compound.

Practical tips:

- *Isotope patterns* are characteristic of compounds and can be used to identify metabolites. For example, for a parent molecule that includes a chlorine atom, an isotope pattern of M, M + 2 with a ratio 3:1 is expected. So, if the chlorine atom is retained in the drug/metabolite, the same pattern is expected (see Table 8.6 for other elements).

- *High resolution MS* can be used to identify the elements present in the molecule and, in many cases, assess the exact nature of the modification. For example, an addition of 14 Da can be due to methylation ($+CH_2$) or oxidation and desaturation ($+O-2H$). In high resolution MS, one is looking at the difference between addition of 14.0266 Da ($12.0107 + (2 \times 1.0079)$ for $+CH_2$) and 13.9841 Da ($15.9994 - (2 \times 1.0079)$ for $O-2H$).

- *In-source fragmentation* forms fragment ions that could lead to the wrong interpretation of the molecular ion. For example, in-source fragmentation of glucuronide metabolites leads to cleavage of glucuronic acid (-176 Da) in the source, resulting in an increased abundance of molecular ions of the aglycan metabolite.

- *Formation of adducts* leads to formation of molecular ions that correspond not to $[M + H]^+$, but to $[M + NH_4]^+$ (addition of 17.0306 amu), $[M + Na]^+$ (addition of 21.9818 amu), and $[M + K]^+$ (addition of 38.0904 amu).

- *Knowing the expected metabolites* from a source helps with predicting the possible modifications. For example, glucuronide metabolites are not expected when using liver microsomes without the necessary cofactors (i.e., uridine diphosphate glucuronic acid).

Double bond equivalent (DBE; known also as degree of unsaturation) is the number of double bonds or rings in a molecule. It is derived in the following way:

$$DBE = 1 + C + N/2 - H/2.$$

6.3.2 MS/MS Data (Product Ion Scans)

Product ion scans involve selecting the ion of interest, the molecular ion in most cases, in the first quadrupole and following a fragmentation in the ion source. The fragment ions are then detected in the mass analyzer. See Table 6.2 for some common neutral loss masses for product ion scans.

TABLE 6.1.   Common Phase I and Phase II biotransformations and the corresponding change in mass of the parent compound in high resolution MS

Biotransformation pathway	Formula	Mass change (amu)
Decarboxylation	$-CO_2$	-43.9898
Deethylation	$-C_2H_4$	-28.0313
Oxidative dechlorination	$+OH-Cl$	-17.9662
Demethylation	$-CH_2$	-14.0157
Desaturation	$-2H$	-2.0157
Oxidative defluorination	$+OH-F$	-1.9957
Oxidative deamination	$+O-NH_3$	-1.0316
Double bond reduction	$+2H$	2.0157
Oxidation and desaturation	$+O-2H$	13.9793
Methylation	$+CH_2$	14.0157
Hydroxylation, oxide	$+O$	15.9949
Hydrolysis	$+H_2O$	18.0106
Hydroxylation and methylation	$+O+CH_2$	30.0106
Dihydrolylation	$+2O$	31.9898
Acetylation	$+COCH_3-H$	42.0106
Glycine conjugation	$+NHCH_2CO_2H-OH$	58.0215
Sulfonate conjugation	$+SO_3$	79.9568
Hydrolylation and sulfonation	$+O+SO_3$	95.9517
Taurine conjugate	$NHCH_2CH_2SO_3H-OH$	107.0041
N-acetylcysteine conjugate (mercapturic acid)	$SCH_2CH(CO_2H)$ $NHCOCH_3-H$	161.0147
Glucuronide conjugate	$+C_6H_8O_6$	176.0321
Glutathione (GSH) conjugate	$+C_{10}H_{17}O_6N_3S-2H$	305.0682

TABLE 6.2. Common neutral loss masses from product ion scans

Mass change (amu)	Species lost
17.0265	NH_3
18.0106	H_2O
20.0062	HF
26.0157	C_2H_2
27.0109	HCN
27.9949, 28.0313	CO, C_2H_4
30.0106, 30.0470, 29.9980	CH_2O, C_2H_6, NO
32.0262	CH_3OH
33.9877	H_2S
35.9767, 37.9737	$H^{35}Cl$, $H^{37}Cl$
42.0106, 42.0470, 42.0344	CH_2CO, C_3H_6, C_2H_4N
43.0058, 43.0422	NHCO, C_2H_5N
43.9898	CO_2
46.0055	HCO_2H
56.0626	C_4H_8
60.0211	$C_2H_4O_2$
63.9619	SO_2
79.9262, 81.9241	$H^{79}Br$, $H^{81}Br$
127.9123	HI

6.3.2.1 The Odd Nitrogen Rule

The odd nitrogen rule is a handy tool in metabolite identification. If a compound has an even number of nitrogen atoms (or none), its molecular ion or fragment ions ($[M + H]^+$ or $[M - H]^-$) will have an odd-numbered mass. If a compound has an odd number of nitrogen atoms, its molecular ion will have an even-numbered mass (see Table 6.3). This rule holds true because a nitrogen atom makes three bonds and has an even atomic mass unit (amu). Carbon and oxygen atoms form an even number of bonds and have even amu values, while a hydrogen atom forms an odd number of bonds (one) and has an odd amu.

TABLE 6.3. The odd nitrogen rule

Ions with nitrogen atoms	Even-electron ions	Odd-electron ions
Even or none	Odd-numbered mass	Even-numbered mass
Odd	Even-numbered mass	Odd-numbered mass

6.3.2.2 Derivatization Techniques

Derivatization is a technique used to modify the potential site of modification with a specific functional group (see Table 6.4). From the perspective of metabolite identification, ideally, derivatization should be a one-step process without purification that can be performed at the bench with limited synthetic chemistry skills.

TABLE 6.4. Common derivatization reactions used to aid in metabolite identification

Moiety	Reaction	Reagent
Carboxylic acid	Esterification	Anhydrous methanol +HCl or diazomethane (TMS diazomethane)
Aldehyde	Reduction to alcohol	$LiAlH_4$ or $NaBH_4$
N-Oxide	Reduction	$TiCl_3$
Amine	Acetylation	Acetic anhydride
Phenol	Acetylation	Acetic anhydride

6.3.2.3 Hydrogen/Deuterium (H/D) Exchange

H/D exchange is a reaction in which exchangeable hydrogen atoms in a compound (or a metabolite) are replaced with deuterium. This is done by performing the reaction/incubation in D_2O instead of H_2O. D_2O does not need to be 100% pure but the percent incorporation is calculated based on $\%D_2O$. Changes in molecular ions and fragment ions that contain exchanged hydrogen atoms are detected by MS. Note that if the samples are run in a mobile phase with H_2O, exchangeable deuterium atoms will most likely be replaced back to hydrogen. To avoid this, a direct infusion can be carried out or D_2O can be used in the mobile phase.

6.4 METABOLITE TESTING IN SAFETY (MIST) CONSIDERATIONS

6.4.1 Metabolites in Safety Testing Guidance

A disconnect is sometimes seen between pharmacokinetic (PK) data and the in vivo pharmacodynamic (PD) behavior of a drug, suggesting that metabolites might be partly responsible for observed efficacy and/or toxicity. In this situation, the exact structure of the metabolite can be obtained and synthesized for subsequent testing in in vitro potency assays. However, it is not common to test metabolites in in vivo toxicology studies because the toxicology species is assumed to be exposed to the metabolite when the parent compound is initially dosed. Nevertheless, it is possible that the exposure to metabolites in humans is higher than that in preclinical toxicology studies. A detailed flow chart has been proposed (see Fig. 6.1) to ensure that sufficient metabolite information is obtained in preclinical settings. This flow chart is based on the 2008 FDA (US Food and Drug Administration Guidance for Industry 2008) and 2009 ICH (ICH Harmonized Tripartite Guideline 2009) guidance and is referred to as the Metabolites in Safety Testing (MIST) guidance.

If a metabolite contributes to <10% of the parent systemic exposure (FDA guidance) or the total drug-related exposure (ICH guidance) at steady state, no further testing is needed. Of course a big difference exists between these two criteria, especially when the parent drug contributes to a relatively small percentage of the total drug-related exposure. In most cases, the ICH guidance is the more practical of the two, but it can be applied correctly only if the total drug-related exposure is known, and this requires human mass balance study using radiolabeled material.

If a metabolite contributes to >10% of the systemic exposure, preclinical testing may be required if exposure to the metabolite at the maximum tolerated dose exceeds that observed in humans at the highest clinical dose. The definitive assessment requires preclinical and clinical studies with radiolabeled material so that the absolute exposure of the metabolites can be quantified. However, radiolabeled material is frequently not available until later in the development stage.

The following experiment (Walker et al. 2009) can assess the liability of a metabolite in the absence of radiolabeled material:

1. Mix 50% plasma at the maximum tolerated dose from one of the preclinical toxicology species with 50% blank human plasma.
2. Mix 50% human plasma at the highest clinical dose with 50% blank plasma from one of the preclinical toxicology species.
3. Extract the drug and metabolite from the plasma samples.
4. Analyze the samples back-to-back, profiling the metabolites with LC-MS/MS.

Since the matrix is the same in both samples, the absolute abundance of the MS signal for each metabolite can be compared. The shortcomings of this approach are that (1) unanticipated metabolites may be missed and (2) determining if a metabolite contributes to >10% of the parent systemic exposure or total drug-related exposure may be difficult because the parent compound and metabolite may have different MS ionization responses.

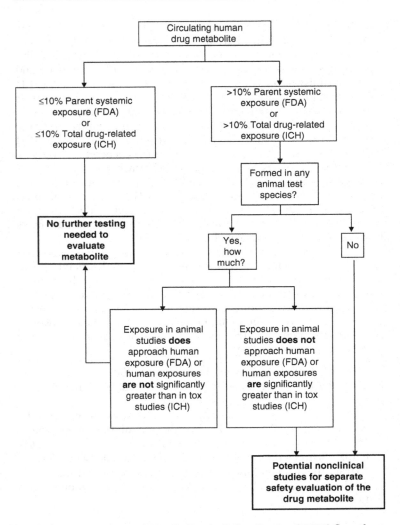

FIGURE 6.1. FDA proposed Metabolites in Safety Testing (MIST) flow chart for assessing sufficient metabolite coverage in human plasma compared to preclinical species.

Monitoring UV response can circumvent the latter disadvantage, but this technique is prone to endogenous interference. Quantitative NMR is another option for monitoring metabolites in the absence of radiolabeled material (Vishwanathan et al. 2009).

AUC pooling is a method of pooling samples from multiple time points to get a single sample with concentrations of the parent drug and its metabolites that represent the total area under the curve (AUC) of the concentration–time profile (Hop et al. 1998).

$$\text{AUC} = C_{\text{pool}}(t_m - t_0)$$

C_{pool} = concentration of pooled sample.
t_m = last timepoint.
t_0 = moment of dosing (usually $t_0 = 0$).
The volumes of the individual aliquots from each timepoint used to create the pooled sample can be determined by:

$$v_0 : v_1 : v_2 \ldots v_j \ldots v_m = k\Delta t_0 : k\Delta t_1 : k\Delta t_2 \ldots k\Delta t_j \ldots k\Delta t_m$$

v = volume of aliquot taken from each timepoint.
k = proportionality constant.

$$\Delta t_0 = t_1 - t_0, \Delta t_1 = t_2 - t_0, \Delta t_2 = t_3 - t_1 \ldots \Delta t_j$$
$$= t_{j+1} - t_{j-1} \ldots \Delta t_m = t_m - t_{m-1}$$

Example: If PK samples were collected for the following time points and $k = 10$, then the following volumes will need to be pooled.

Timepoint (in h)	0	0.25	0.5	1	2	4	8	24
Δt		0.25	0.5	0.75	1.5	4	6	20	16
v		10	20	30	60	160	240	800	640

AUC pooling can reduce the number of samples for bioanalysis from PK studies. This methodology may be even more valuable for metabolism studies to determine the relative amount of the parent drug and its metabolites that an animal is exposed to.

The following caveats have to be taken into consideration when interpreting the MIST guidance:

- Studies should preferably be done at steady state because the half-lives of metabolites may be different from that of the parent compound.
- Pharmacological and toxicological behavior is likely to be dependent on free plasma exposure, and, therefore, differences in plasma protein binding between the parent compound and metabolites should be considered.
- The guidance does not require both preclinical toxicology species to form the metabolite at adequate exposures.
- *"Some metabolites are not of toxicological concern (e.g., most glutathione conjugates) and do not warrant testing."* (ICH Harmonised Tripartite Guideline 2009).
- Even if the preclinical exposure of a metabolite is less than that encountered in the clinic, preclinical safety studies with this metabolite may not be realistic or practical (e.g., glucuronides).
- *"The nonclinical characterization of metabolites with an identified cause for concern (e.g., a unique human metabolite) should be considered on a case-by-case basis."* (ICH Harmonized Tripartite Guideline 2009).
- *"This guidance does not apply to some cancer therapies where a risk–benefit assessment is considered."* (US Food and Drug Administration Guidance for Industry 2008).

6.5   OVERVIEW OF BIOACTIVATION

Drug attrition during the development stage is high and puts a strain on the pharmaceutical industry. Preclinical and clinical toxicity is responsible for about one-third of the attrition of compounds in development, and this number has been gradually increasing (Kola and Landis 2004). Initiatives have been taken by industry-wide organizations such as PhRMA to address this issue, and new studies, for example the recently published research on finding more-predictive markers for kidney toxicity (Ozer et al. 2010), may assist this situation.

Reactive metabolites have been implicated in some cases of toxicity, but the link between reactive metabolites and toxicity is complex and in most cases difficult to establish. The liver is frequently the target organ of reactive metabolites because of the high levels of drug and drug metabolizing enzymes in the liver. A well-established example is the severe hepatoxicity associated with bioactivation of acetaminophen to *N*-acetyl-*p*-quinoneimine (NAPQI) at high doses. After saturating the detoxification conjugative pathways and depleting intracellular glutathione (GSH), NAPQI reacts with critical cellular protein nucleophiles and produces cellular necrosis (see Fig. 6.2).

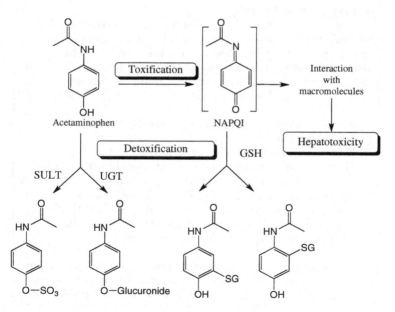

FIGURE 6.2. Cytochrome P450-mediated bioactivation of acetaminophen.

In other cases, any observed toxicity may be rare and idiosyncratic and can lead to disorders such as a severe rash. The problem with idiosyncratic toxicity is that, by definition, it is rare and unpredictable and, therefore, may escape observation in clinical trials. In addition, idiosyncratic toxicity does not appear to be dose dependent. Finally, some reactive metabolites react with genetic material.

For all the above reasons, most pharmaceutical companies have established assays to determine the bioactivation of drug candidates. Reactive metabolites are usually short lived and, therefore, difficult to detect directly. Thus, assays that monitor surrogate endpoints have been employed:

- Trapping reactive metabolites with GSH, cyanide, methoxylamine, semicarbazide, or other nucleophiles.
- Measuring the extent of covalent modification of proteins.
- Determining the time-dependent inhibition of cytochrome P450 (P450) enzymes.
- Monitoring the formation of stable metabolites that points to formation of reactive metabolites, such as the detection of carboxylic acids from terminal acetylenes.

Idiosyncratic Toxicity

An idiosyncratic drug reaction is defined as hypersensitivity to a substance without a direct connection to the pharmacological target of the drug. Although the underlying mechanism is usually unknown, the toxicological response appears to involve direct hepatotoxicity and/or adverse immune reactions. Characteristics of an idiosyncratic drug reaction are as follows:

1. A low incidence (e.g., bromfenac toxicity is seen in 1 out of 20,000 patients).
2. No uniform response from patient to patient.
3. No clear dose/exposure–response relationship. The response is usually more pronounced after repetition of the treatment. An example is Stevens–Johnson syndrome caused by carbamazepine. Note that some patients may be genetically predisposed to such drug reactions, such as those that might carry a specific human leukocyte antigen.

"It is now assumed that most idiosyncratic drug reactions are due to reactive metabolites, and yet most drugs form reactive metabolites to some degree, and we can not predict with

---

Continued

> any degree of certainty which drugs will be associated with a high incidence of idiosyncratic reactions". (Uetrecht 2002)

6.5.1 Trapping of Reactive Metabolites

As a result of advances in LC-MS/MS technology, detecting adducts with nucleophiles, such as GSH, cyanide, methoxylamine, and semicarbazide, is practical (see Fig. 6.3).

6.5.1.1 GSH Conjugates

GSH conjugates are suited for detection of soft electrophiles such as Michael acceptors and epoxides. To screen for GSH conjugates use the following:

- A 129 Da (glutamic acid) neutral loss scan in the positive ion mode on a triple quadrupole mass spectrometer.
- A m/z 272 precursor ion scan in the negative ion mode on a triple quadrupole mass spectrometer (Dieckhaus et al. 2005). This method is described as having low interference.

Detection of GSH conjugates is more selective when using a mixture of regular and stable-labeled GSH (Yan and Caldwell 2004).

6.5.1.2 GSH Analogs

GSH detection by LC-MS/MS is not quantitative, and other trapping analogs, such as [^{3}H]GSH (Thompson et al. 1995) and dansyl-GSH (Gan et al. 2009), are used for this purpose. The limitation of these studies is that the trapped GSH conjugates usually form a small portion of the total GSH conjugates.

In addition, quaternary ammonium GSH analogs produce more-sensitive GSH adducts (Soglia et al. 2006). Note that analogs of GSH are not catalyzed by glutathione S-transferase (GST).

Relationship Between Thiol Adducts and Toxicity

A relationship is proposed to exist between the estimated total human in vivo thiol adduct burden (normalized for the daily dose) and drug-induced toxicity (Gan et al. 2009). In this study, dansyl-GSH was used to facilitate quantitation of adducts.

6.5.1.3 Cyanide Conjugates

Cyanide conjugates are suited for detection of iminium ions. To screen for cyanide conjugates use the following:

- A 27 Da neutral loss scan in the positive ion mode on a triple quadrupole mass spectrometer (Argoti et al. 2005). A mixture of CN^- and stable-labeled $^{13}C^{15}N^-$ enhances the selectivity for detection of cyanide conjugates.

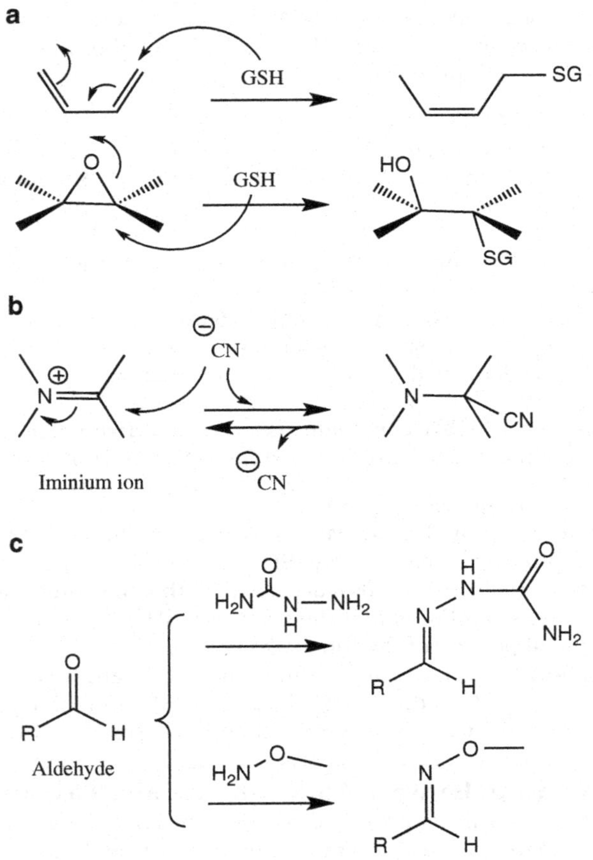

FIGURE 6.3. Mechanisms of trapping reactive intermediates by glutathione (GSH), cyanide (CN^-), methoxylamine, and semicarbazide.

6.5.1.4 Methoxylamine and Semicarbazide Conjugates

These conjugates are suited for trapping aldehydes. No straightforward fragmentation pattern exists for MS/MS methods in the scan mode for automated data acquisition.

6.5.2   Covalent Modification of Proteins

Determination of covalent binding of drugs to endogenous biomolecules, such as proteins, has been conducted for many years, and high-throughput methods have been established (Day et al. 2005). Some companies have integrated covalent binding assays into the drug discovery stage and, although no fixed cut-off values are broadly applicable, a value of >50 pmol/mg protein is proposed as the threshold beyond which the drug candidate should not be pursued further (Evans et al. 2004). When multiple radiolabeled compounds are available, structure–activity relationships can be evaluated, which can lead to development of drugs that reduce or eliminate the extent of covalent binding. However, no apparent quantitative link exists between covalent binding and toxicity (Obach et al. 2008).

6.5.3   Time-Dependent Inhibition of P450 Enzymes

Reactive metabolites can inactivate P450 enzymes by covalently modifying the heme group or the apoprotein of the enzyme. This topic is covered in greater detail in Chap. 5.

6.5.4   Bioactivation in the Context of Drug Discovery and Development

The steps a metabolism scientist should take when bioactivation (defined by formation of GSH or other adducts, covalent binding, or mechanism-based inactivation of P450 enzymes) is observed continues to be debated in the literature. It is clear that covalent binding can be a liability, but it is not necessarily a showstopper. The following aspects should be taken into consideration:

- The extent of adduct formation, covalent binding, or mechanism-based P450 enzyme inactivation.
- Not all in vitro observations translate into in vivo liabilities because alternative, detoxifying clearance pathways may exist in vivo. For example, in the case of raloxifene, in vitro incubation with human liver microsomes and nicotinamide adenine dinucleotide phosphate (NADPH) leads to formation of GSH conjugates and covalent binding. However, the main route of elimination for raloxifene is glucuronidation in the intestine and liver (Dalvie et al. 2008).
- A low dose and/or low systemic exposure greatly reduces the risk for serious toxicity.
- Duration of therapy – short term treatment (e.g., antibiotics) versus long term use in relatively healthy patients.
- Therapeutic area – life-threatening diseases versus drug lifestyle drugs (e.g., obesity).

- Target population – some patient populations may be more sensitive to bioactivation than others.
- If a prototype for a new target or compound aims for best in class classification.
- Species differences – not all species process drugs in the same way, and bioactivation may be unique to one species.

Nevertheless, bioactivation studies can provide valuable mechanistic insights that can be incorporated in the overall optimization of a drug structure.

Efavirenz, a non-nucleoside reverse transcriptase inhibitor for the treatment of HIV, presents a classic example of the role of bioactivation in drug development (Mutlib et al. 2000). Efavirenz causes severe renal tubular renal cell necrosis in rats, but not in monkeys, and scientists were greatly concerned that this condition would manifest itself in humans as well. However, detailed metabolism studies showed that the differences in toxicity between the species are due to differences in the production and processing of reactive metabolites. Rats produce a unique GSH adduct that hydrolyzes to a cysteineglycine conjugate that ultimately leads to toxicity (see Fig. 6.4). Indeed, in this case human metabolism is more like that in monkeys, and no renal toxicity was observed in humans.

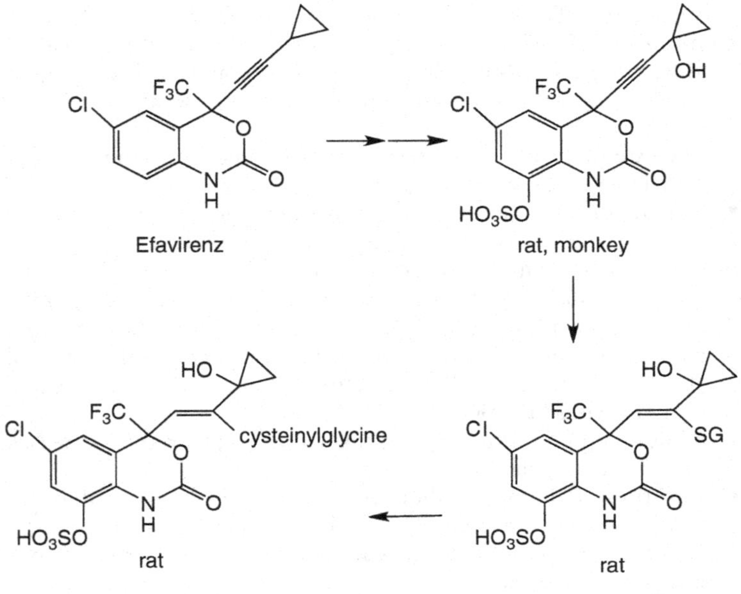

Figure 6.4. Metabolism of efavirenz in rats, monkeys, and humans.

Finally, the view towards bioactivation is also influenced by the ease of addressing it. If multiple lead series are pursued and one or more series do not display bioactivation, these leads should be given serious consideration for further development. The most pragmatic approach may be to simply avoid compounds with certain structural characteristics that are associated with metabolic bioactivation (Kalgutkar and Soglia 2005).

6.6 BIOTRANSFORMATION/BIOACTIVATION OF MOIETIES

Here, we describe major biotransformation pathways by Phase I and Phase II DMEs (Figs. 6.5–6.15).

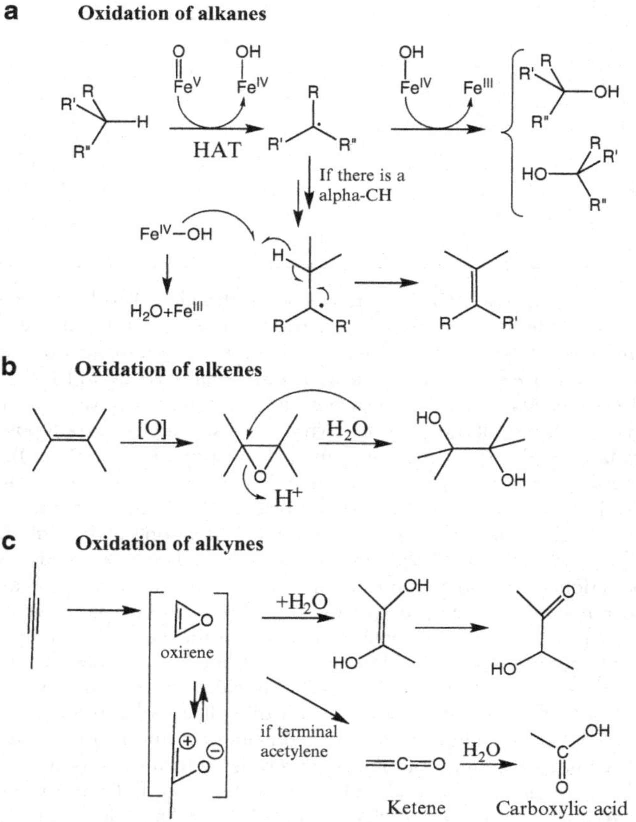

a   Oxidation of alkanes

b   Oxidation of alkenes

c   Oxidation of alkynes

FIGURE 6.5. **Oxidation of alkanes, alkenes, and alkynes.** *Alkanes.* P450-mediated oxidation of carbon atoms proceeds via hydrogen atom transfer

FIGURE 6.5 *Continued* (HAT; homolytic bond breakage) to form a carbon radical intermediate, followed by a hydroxyl radical rebound reaction. *Selectivity of the site of oxidation.* When the carbon radical intermediate mentioned above is stabilized, the activation energy required for the reaction to proceed is lowered. The carbon radical is usually most stable in the benzylic position, followed by the tertiary and secondary carbon positions (benzylic > allylic (branch > unbranched) > aliphatic (3 > 2 > 1)). This relationship, however, does not always hold true since other factors are involved in oxidation, such as binding orientation to the enzyme and the propensity of neighboring moieties to induce oxidation (allylic moieties for example). *Alkenes.* P450-mediated oxidation of alkenes proceeds via the introduction of an oxygen atom to form an epoxide. Most epoxides are not stable, but some epoxides, such as the 10,11-epoxide of carbamazepine, are stable. The stability of the epoxide depends on the electron density of the double bond; the higher the electron density, the more stable the epoxide. This can be followed by enzymatic (via epoxide hydrolases) or non-enzymatic opening of the epoxide to form a diol. The diol can loose a water molecule to form a ketone or aldehyde (via an enol intermediate). *Alkynes.* P450-mediated oxidation of alkynes proceeds by formation of an oxirene, which rearranges to a ketone and reacts with nucleophiles to form esters and amides, or with water to give an acid. Terminal acetylenes can also be oxidized to ketoradicals, which inactivate the P450 enzyme by binding to the heme nitrogen.

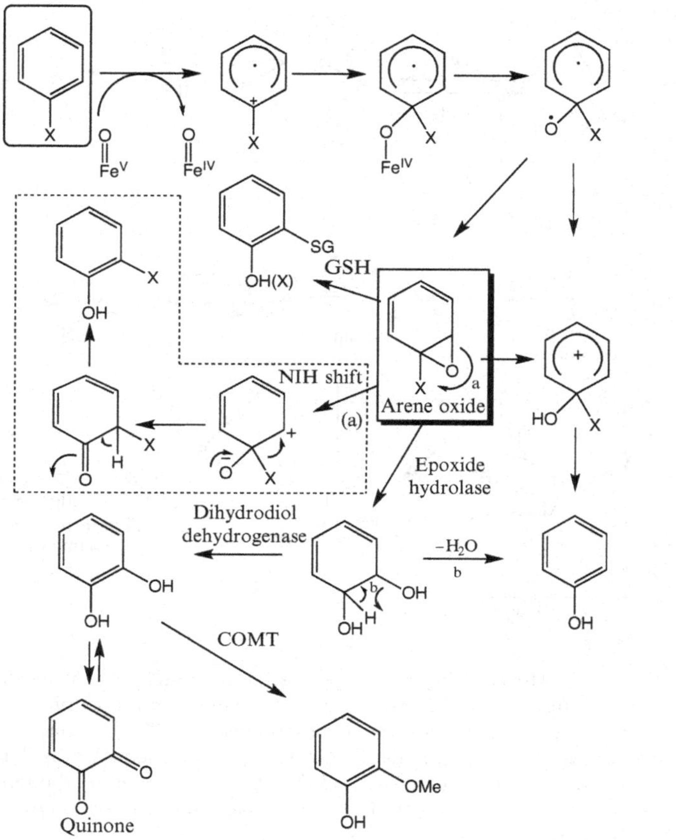

FIGURE 6.6. **Aromatic ring oxidation.** Several mechanisms have been proposed for the oxidation of aromatic rings. One mechanism begins with a SET, which results in the formation of a cation radical of the benzene ring. This step is followed by nucleophilic attack of Fe^{IV}–O^-. The other mechanism involves formation of an epoxide, similar to what was described for alkene oxidation. The arene oxide formed can rearrange in two ways. In one mechanism, an NIH shift is observed in which X migrates to the neighboring carbon (X can be CH_3, Cl, Br, F). The arene oxide can also react with GSH (catalyzed by GST) or water (catalyzed by microsomal expoxide hydrolase).

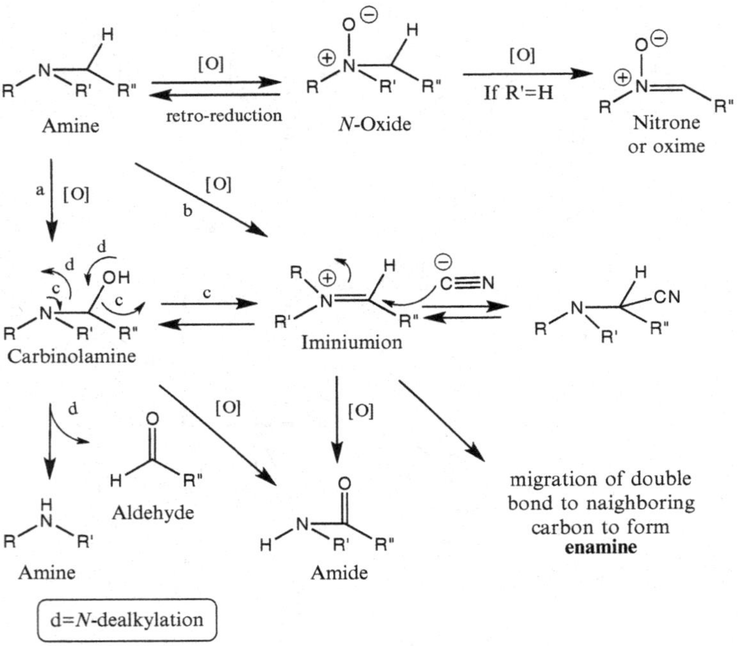

FIGURE 6.7. **Aliphatic nitrogen metabolism.** Amines can be oxidized to several different types of metabolites, including a carbinolamine intermediate that leads to N-dealkylated compounds, N-oxides, nitrones, oximes, iminium ions, amides, and enamines. For P450-mediated oxidation of amines and amides, the first step of the reaction is generally thought to be via hydrogen atom transfer (HAT) from the alpha carbon followed by a hydroxyl radical rebound.

An NIH shift is a chemical rearrangement in which a chemical group undergoes intramolecular migration during the process of oxidation (see Fig. 6.6).

Ether oxidation

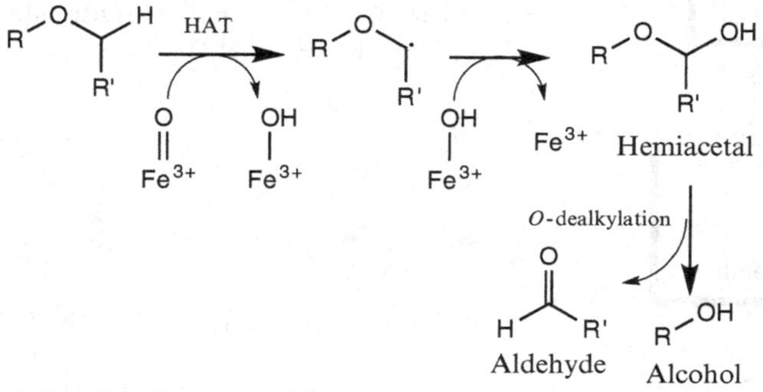

Methylenedioxybenzene oxidation

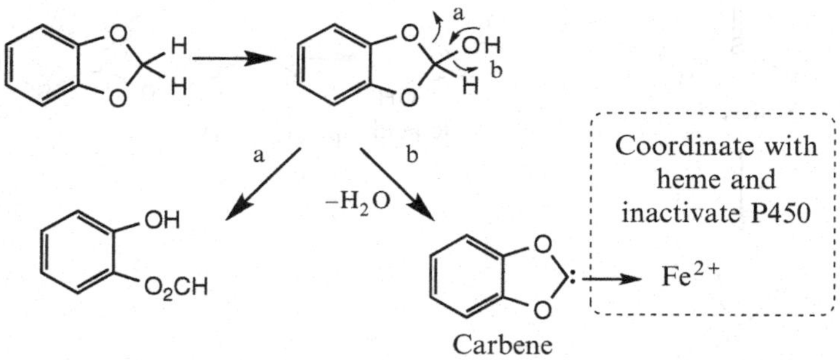

FIGURE 6.8. **Metabolism of oxygen-containing compounds.** *Ethers.* The P450-mediated oxidation of ethers proceeds by oxidation of the alpha carbon via HAT to form a radical, followed by a hydroxyl radical rebound. This mechanism results in the formation of a hemiacetal that hydrolyses to an alcohol and an aldehyde product. *Methylenedioxybenzenes.* The P450-mediated oxidation of methylenedioxybenzenes proceeds via oxidization of the methylene to a carbene (a carbon with six valence electrons) intermediate. This intermediate coordinates with the heme in the P450 active site and results in inactivation of the P450 enzyme.

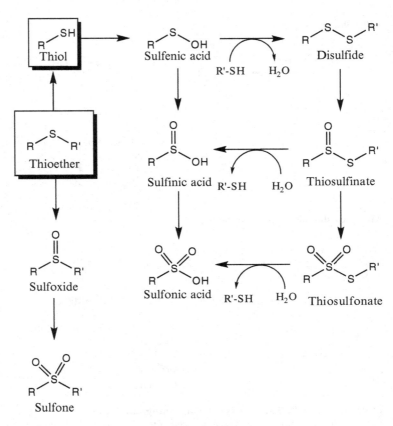

FIGURE 6.9. **Metabolism of thiols and thioethers.** Thioethers are oxidized to S-dealkylation products (including a thiol as one of the products). Both thioethers and thiols can be oxidized to form a number of metabolites. The sulfur can also be oxidized to a sulfoxide or a sulfone.

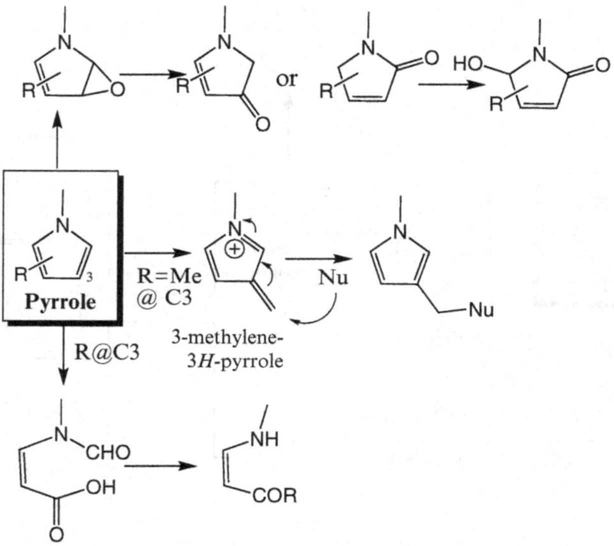

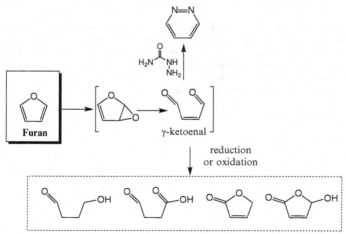

FIGURE 6.10. **Pyrrole oxidation.** Pyrrole is oxidized via epoxidation, leading to various oxidative metabolites. In the case of 3-methylpyrrole (or 3-methyl-indole), oxidation can proceed via HAT from the methyl group, leading to formation of a 3-methylene-3H-pyrrole intermediate. This intermediate can react with nucleophiles such as GSH to form a stable conjugate.

FIGURE 6.11. **Furan oxidation.** P450-mediated furan oxidation proceeds by epoxidation and leads to formation of a γ-ketoenal. This reactive intermediate can be trapped by semicarbazide. It can also rearrange and/or undergo further oxidation or reduction as shown.

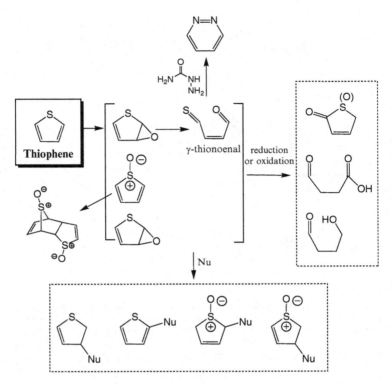

FIGURE 6.12. **Thiophene oxidation.** The three proposed mechanisms of thiophene oxidation are: epoxidation, S-oxide formation, and S–Cl formation (no direct evidence of an S–Cl intermediate). Semicarbazide traps the epoxide intermediate to form pyridazine, which suggests the formation of a γ-thioenal intermediate. Thiophenes can also rearrange and/or be oxidized or reduced to various stable metabolites.

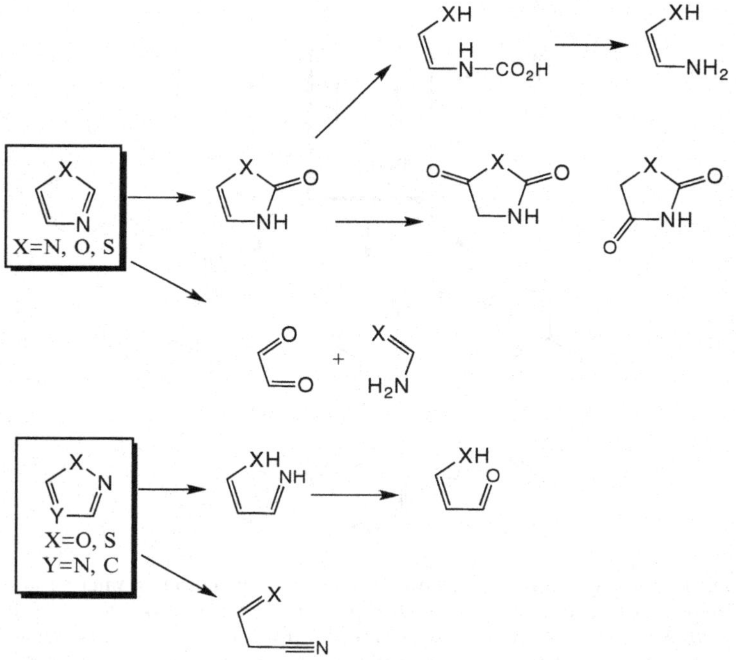

FIGURE 6.13. **Oxidation of imidazole, oxazole, and thioazoles.** Oxidation of 1,3 analogs can lead to a number of oxidative metabolites that include open ring systems. The 1,2 analogs can be reduced to form an open ring by enzymes such as AO (see Chap. 2).

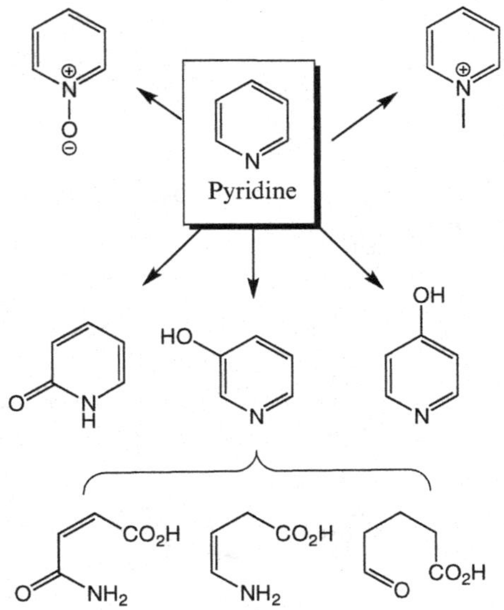

FIGURE 6.14. **Pyridine oxidation.** Besides ring oxidation, similar to aromatic ring oxidation (see Sect. 6.6.2), the nitrogen can also be oxidized to form *N*-oxide or *N*-methylated products. Multistep oxidations and reductions can produce an open ring product. Here, we present a few examples, but many more have been reported after undergoing reactions with microbes under anaerobic and aerobic conditions (Kaiser et al. 1996).

Reaction	Biotransformation	Δ mass
Glucuronidation	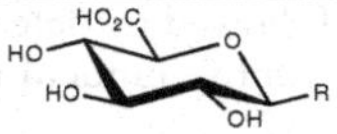	+176.0321
Sulfonation	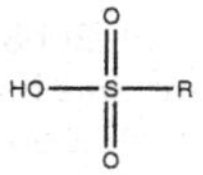	+79.9568
GSH conjugation	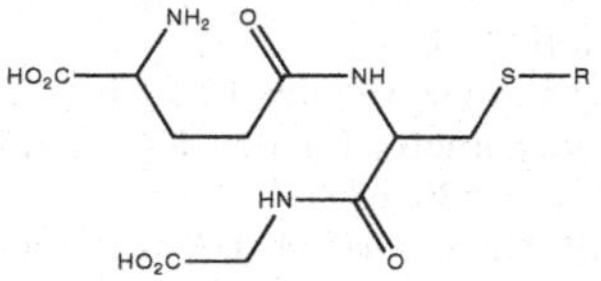	+305.0682
Acetylation	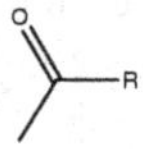	+42.0106
Methylation		+14.0157
Glycine conjugation	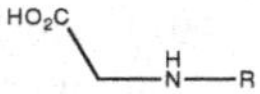	+73.0164
Taurine conjugation		+122.9990

FIGURE 6.15. Common conjugative reactions.

References

Argoti D, Liang L, Conteh A et al (2005) Cyanide trapping of iminium ion reactive intermediates followed by detection and structure identification using liquid chromatography–tandem mass spectrometry (LC-MS/MS). Chem Res Toxicol 18:1537

Dalvie D, Kang P, Zientek M et al (2008) Effect of intestinal glucuronidation in limiting hepatic exposure and bioactivation of raloxifene in humans and rats. Chem Res Toxicol 21:2260–2271

Day SH, Mao A, White R et al (2005) A semi-automated method for measuring the potential for protein covalent binding in drug discovery. J Pharmacol Toxicol Meth 52:278–285

Dieckhaus CM, Fernandez-Metzler CL, King R et al (2005) Negative ion tandem mass spectrometry for the detection of glutathione conjugates. Chem Res Toxicol 18:630–638

Evans DC, Watt AP, Nicoll-Griffith DA et al (2004) Drug–protein adducts: An industry perspective on minimizing the potential for drug bioactivation in drug discovery and development. Chem Res Toxicol 17:3–16

FDA (2008) US Food and Drug Administration guidance for industry: safety testing of drug metabolites. www.fda.gov/downloads/Drugs/ GuidanceCompliance RegulatoryInformation/Guidances/ucm079266. pdf. Accessed 28 June 2010

Frederick CB, Obach RS (2010) Metabolites in safety testing: "MIST" for the clinical pharmacologist. Clin Pharmacol Ther 87:345–350

Gan J, Ran Q, He B et al (2009) In vitro screening of 50 highly prescribed drugs for thiol adduct formation: comparison of potential for drug-induced toxicity and extent of adduct formation. Chem Res Toxicol 22:690–698

Hop CECA, Wang Z, Chen Q et al (1998) Plasma-pooling methods to increase throughput for in vivo pharmacokinetic screening. J Pharm Sci 87:901–903

ICH Expert Working Group (2009) ICH harmonised tripartite guideline: guidance on nonclinical safety studies for the conduct of human clinical trials and marketing authorization for pharmaceuticals M3(R2). http:// www.ich.org/LOB/media/MEDIA5544.pdf. Accessed 28 June 2010

Ju C, Uetrecht JP (2002) Mechanism of idiosyncratic drug reactions: reactive metabolites formation, protein binding and the regulation of the immune system. Curr Drug Metab 3:367–377

Kalgutkar AS, Soglia JR (2005) Minimising the potential for metabolic activation in drug discovery. Expert Opin Drug Metab Toxicol 1:91–142

Kaiser J-P, Feng Y, Bollag J-M (1996) Microbial metabolism of pyridine, quinoline, acridine, and their derivatives under aerobic and anaerobic conditions. Microbiol Rev 60:483–498

Kola I, Landis J (2004) Can the pharmaceutical industry reduce attrition rates? Nat Rev Drug Disc 3:711–715

Liu DQ, Hop CECA (2005) Strategies for characterization of drug metabolites using liquid chromatography–tandem mass spectrometry in conjunction with chemical derivatization and on-line H/D exchange approaches. J Pharm Biomed Anal 37:1–18

Mutlib AE, Gerson RJ, Meunier PC et al (2000) The species-dependent metabolism of efavirenz produces a nephrotoxic glutathione conjugate in rats. Toxicol Appl Pharmacol 169:102–113

Obach RS, Kalgutkar AS, Soglia JR et al (2008) Can in vitro metabolism-dependent covalent binding data in liver microsomes distinguish hepatotoxic from nonhepatotoxic drugs? An analysis of 18 drugs with consideration of intrinsic clearance and daily dose. Chem Res Toxicol 21:1814–1822

Ozer JS, Dieterle F, Troth S et al (2010) A panel of urinary biomarkers to monito reversibility of renal injury and a serum marker with improved potential to assess renal function. Nat Biotechnol 28:486–494

Roberts KM, Jones JP (2010) Anilinic N-oxides support cytochrome P450-mediated N-dealkylation through hydrogen-atom transfer. Chemistry 16:8096–8107

Soglia JR, Contillo LG, Kalgutkar AS et al (2006) A semiquantitative method for the determination of reactive metabolite conjugate levels in vitro utilizing liquid chromatography–tandem mass spectrometry and novel quaternary ammonium glutathione analogues. Chem Res Toxicol 19: 480–490

Testa B, Caldwell J (1994) The metabolism of drugs and other xenobiotics. Biochemistry of redox reactions (metabolism of drugs and other xenobiotics). Academic Press, San Diego, CA

Thompson DC, Perera K, London R (1995) Quinone methide formation from para isomers of methylphenol (cresol), ethylphenol, and isopropylphenol: relationship to toxicity. Chem Res Toxicol 8:55–60

Uetrecht JP (2002) Preface. Curr Drug Metab 3:i–i(1)

Vishwanathan K, Babalola K, Wang J et al (2009) Obtaining exposures of metabolites in preclinical species through plasma pooling and quantitative NMR: addressing metabolites in Safety Testing (MIST) guidance without using radiolabeled compounds and chemically synthesized metabolite standards. Chem Res Toxicol 22:311–322

Walker D, Brady J, Dalvie D et al (2009) A holistic strategy for characterizing the safety of metabolites through drug discovery and development. Chem Res Toxicol 22:1653–1662

Yan Z, Caldwell GW (2004) Stable-isotope trapping and high-throughput screenings of reactive metabolites using the isotope MS signature. Anal Chem 76:6835–6847

Additional Reading
See Chap. 2

Chapter 7
Prediction of Human Pharmacokinetics

Abstract
The prediction of human pharmacokinetics is an extremely difficult endeavor during the selection of drug candidates for further
human clinical testing. Despite a variety of available in vitro and
in vivo methodologies, successful predictions are still difficult
when performing them prospectively. This chapter gives a general
overview of in vitro and in vivo methodologies used to predict
human pharmacokinetics.

Contents

7.1 LIST OF ABBREVIATIONS

ADME	Absorption, distribution, metabolism, and excretion
BrW	Brain weight
$CL_{hepatic}$	Hepatic clearance
CL_{int}	Intrinsic clearance
CL_u	Unbound clearance
F	Bioavailability
F_a	Fraction absorbed from the intestine

S.C. Khojasteh et al., *Drug Metabolism and Pharmacokinetics
Quick Guide*, DOI 10.1007/978-1-4419-5629-3_7,
© Springer Science+Business Media, LLC 2011

F_g	Fraction that escapes intestinal metabolism
F_h	Fraction that escapes hepatic metabolism
f_u	Unbound fraction in blood/plasma
f_{umic}	Unbound fraction in microsomes
f_{ut}	Unbound fraction in tissues
HPGL	Hepatocytes per gram of liver
IVIVE	In vitro–in vivo extrapolation
K_m	Michaelis–Menten constant (i.e., substrate concentration when v is ½ of V_{max})
MLP	Maximum life span
MPPGL	Microsomal protein per gram of liver
MRT	Mean residence time
P450	Cytochrome P450
PBPK	Physiologically based pharmacokinetic model
PK	Pharmacokinetic
$P_{microsome}$	Amount of microsomal protein in the incubation
Q	Hepatic blood flow
R_e/I	Ratio of binding proteins in extracellular fluid (except plasma) to binding proteins in plasma
$t_{1/2(in\ vitro)}$	In vitro half-life
V_d	Volume of distribution
$V_{d,u}$	Unbound volume of distribution
V_e	Extracellular fluid volume
$V_{incubation}$	Incubation volume
V_{max}	Maximum rate of the metabolic reaction
V_p	Plasma volume
V_r	"Remainder" of the fluid volume

7.2 BASIC CONCEPTS

One of the key functions of DMPK scientists in drug discovery is prediction of human pharmacokinetics and the human dose. The general concept is outlined in Fig. 7.1. The three most relevant parameters are bioavailability (F), clearance (CL; clearance also contributes to F), and volume of distribution (V_d). As described in Chap. 3, F is determined by the fraction absorbed from the intestine (F_a), the fraction that escapes intestinal metabolism (F_g), and the fraction that escapes hepatic metabolism (F_h). Other pharmacokinetic (PK) parameters, such as half-life ($t_{1/2}$), can be derived from these parameters. Numerous methods are available for prediction of these parameters, and they are listed in Table 7.1. The most commonly used methods are described in subsequent sections. More sophisticated predictions can be made with a physiologically based pharmacokinetic (PBPK) model using a range of preclinical in vitro and/or in vivo data.

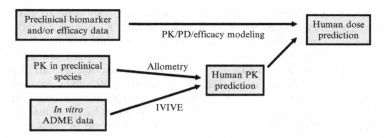

FIGURE 7.1. Flowchart depicting methodology for human pharmacokinetic and dose prediction. ADME = absorption, distribution, metabolism, excretion; IVIVE = In vitro–in vivo extrapolation; PD = pharmacodynamic; PK = pharmacokinetic.

TABLE 7.1. Most commonly used methods to predict human pharmacokinetics

Pharmacokinetic parameter	Methods
Fraction absorbed	Solubility and dissolution data
	In vitro permeability data from Caco-2 or MDCK cell or PAMPA studies
	F_a from preclinical PK studies
Clearance	In vitro–in vivo extrapolation using recombinant P450, microsome or hepatocyte data (with or without free fraction correction)
	Allometry (with or without the rule of exponents; with or without free fraction correction)
	Single species (allometric) scaling (with or without free fraction correction)
	Single species liver blood flow method
	Tang and Mayersohn method
Volume of distribution	Allometry (with or without free fraction correction)
	Single species scaling (with or without free fraction correction)
	Oie–Tozer method

7.3   PREDICTION OF HUMAN FRACTION ABSORBED

The fraction absorbed is influenced by the intestinal solubility of the drug and the permeability across enterocytes. Solubility and dissolution rate studies can predict if absorption is limited by the solubility of the drug, and this is reflected by the maximum absorbable dose. Note that the magnitude of the anticipated human dose should be taken into consideration as well. In vitro permeability studies involving Caco-2 or MDCK cells or PAMPA provide a good idea about the intrinsic permeability of the drug. Details are provided in Chap. 4. Efflux by transporters can limit the fraction absorbed, but intestinal transporters can be saturated relatively easy. Finally, preclinical PK data can be used to predict F_a and the first-order absorption rate constant (k_a). The fraction absorbed combined with knowledge of intestinal metabolism and systemic clearance can be used to predict F.

Preclinical Models for Human Drug Absorption

Although monkeys may be good models to predict F_h in humans, $F_a \times F_g$ is frequently substantially smaller in monkeys than in humans for drugs that undergo a significant degree of metabolism (Akabane et al. 2010). This probably reflects an increased capacity for intestinal metabolism in monkeys, because an earlier study showed that F_a in monkeys correlates well with F_a in humans (Chiou and Buehler 2002).

Although dogs are commonly used to study oral absorption, F_a in dogs is frequently larger than F_a in humans (Chiou et al. 2000). In addition, T_{max} tends to be longer in humans than in dogs. The correlation between F_a in rats and humans is more robust (Chiou and Barve 1998).

7.4 PREDICTION OF HUMAN CLEARANCE

Clearance predictions can be based on either human in vitro ADME data or in vivo PK data from preclinical studies.

7.4.1 In Vitro–In Vivo Extrapolation

In vitro–in vivo extrapolation is the process by which organ clearance is scaled up using in vitro data. Since the liver is the main organ involved in metabolism of xenobiotics, this section focuses on the liver as the organ of interest. A flowchart of the process of in vitro–in vivo extrapolation is shown in Fig. 7.2.

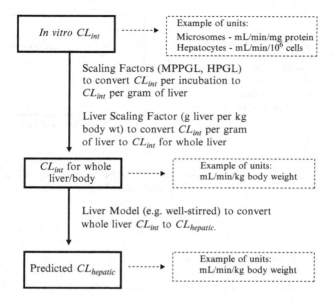

FIGURE 7.2. Flowchart depicting the process of in vitro–in vivo extrapolation. $CL_{hepatic}$ = hepatic clearance; CL_{int} = intrinsic clearance; HPGL = hepatocytes per gram of liver; MPPGL = microsomal protein per gram of liver.

As mentioned above, we will be covering only $CL_{hepatic}$ in this chapter. For drugs and compounds that are eliminated via organs other than the liver, estimated organ clearances can be summed together to get an estimate of total body clearance.

7.4.2 In Vitro Methods of Determining Intrinsic Clearance

Metabolic CL_{int} is a measure of the ability of hepatocytes to eliminate a drug or compound irrespective of other external factors, such as protein binding and hepatic blood flow. Determination of metabolic CL_{int} can be accomplished using traditional enzyme kinetic methodologies or substrate depletion methods.

7.4.2.1 Determination of Intrinsic Clearance Using Michaelis–Menten Kinetic Parameters

The relationship between the rate of a metabolic reaction and the substrate concentration is depicted in Fig. 7.3.

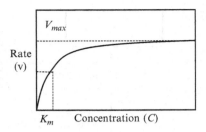

FIGURE 7.3. Plot of relationship between metabolic reaction rate (v) and substrate concentration (C). K_m = Michaelis–Menten constant; V_{max} = maximum rate of the metabolic reaction.

The Michaelis–Menten equation describes this relationship for many metabolic reactions and is presented below:

$$v = \frac{V_{max} \times C}{K_m + C} \tag{7.1}$$

v = rate of the metabolic reaction.
V_{max} = maximum rate of the metabolic reaction.
K_m = Michaelis–Menten constant (i.e., substrate concentration when v is ½ of V_{max}).
C = concentration of the substrate (i.e., drug or compound of interest).

Michaelis–Menten kinetic parameters can be determined using in vitro data under conditions of linearity with respect to incubation time and either microsomal protein concentrations (for microsomal incubations) or number of cells (for hepatocyte incubations). Once Michaelis–Menten kinetic parameters are estimated, the metabolic CL_{int} can be calculated as follows:

$$CL_{int} = \frac{V_{max}}{K_m + C} \text{ or}$$
$$CL_{int} = \frac{V_{max}}{K_m} \text{ under linear conditions where } C << K_m \tag{7.2}$$

Based on the above equations, CL_{int} is concentration dependent at high substrate concentrations approaching K_m. For most drugs, the concentrations administered in vivo are under conditions of linearity with respect to CL_{int}. CL_{int} estimated from in vitro incubations are in units of volume/time/mg of microsomal protein for microsomes (e.g., µL/min/mg microsomal protein) and volume/time/number of cells for hepatocytes (e.g., µL/min/10^6 cells).

Determination of CL_{int} from Michaelis–Menten kinetic parameters can be labor intensive. As can be deduced from Fig. 7.3, in vitro incubations must be performed at multiple substrate concentrations for a good estimate of V_{max} and K_m. In addition, the metabolite reaction that is being monitored must be the formation of a major metabolite for a primary metabolic pathway in order for a prediction to be accurate. Alternatively, for cases in which multiple major metabolic pathways are responsible for drug elimination, the formation rates of the primary metabolites from these multiple pathways must be monitored, and CL_{int} of each pathway must be estimated and summed together to get the overall CL_{int}. The substrate depletion rate may also be estimated at a range of substrate concentrations to estimate a hybrid overall apparent V_{max} and K_m. Regardless, every method utilizing Michaelis–Menten kinetic parameters requires that incubations be performed at a wide range of substrate concentrations. Due to the more resource intensive nature of this method of CL_{int} estimation, substrate depletion estimation methods performed at one substrate concentration with $C << K_m$ have become common.

7.4.2.2 Determination of Intrinsic Clearance Using Substrate Depletion Method at a Single Substrate Concentration (In Vitro Half-Life Method)

Metabolic CL_{int} can be estimated using the in vitro $t_{1/2}$ of a microsome or hepatocyte incubation. Using this method, the substrate concentration must be less than K_m. The following equation is used to estimate CL_{int} from the in vitro $t_{1/2}$:

$$CL_{int} = \frac{0.693}{t_{1/2(\text{in vitro})}} \left(\frac{V_{\text{incubation}}}{P_{\text{microsome}}} \right) \tag{7.3}$$

$t_{1/2(\text{in vitro})}$ = in vitro half-life
$V_{\text{incubation}}$ = volume of the incubation
$P_{\text{microsome}}$ = amount of microsomal protein in the incubation

If $t_{1/2(\text{in vitro})}$ is in minutes, $V_{\text{incubation}}$ is in μL, and $P_{\text{microsome}}$ is in mg, then CL_{int} is in units of μL/min/mg microsomal protein. It is important to keep track of units when performing in vitro–in vivo extrapolation.

The in vitro $t_{1/2}$ method can also be applied to hepatocytes, in which case $P_{\text{microsome}}$ would be replaced with "the number of hepatocytes in the incubation". Since the number of hepatocytes in an incubation is usually expressed as "X $\times$ 10^6 cells", the units for CL_{int} in the preceding example would be $\mu L/min/10^6$ cells.

A common concentration used for estimation of CL_{int} using the in vitro $t_{1/2}$ method is 1 μM.

7.4.3 Scaling Factors for In Vivo–In Vitro Extrapolation of Intrinsic Clearance

In this section, we present scaling factors for the conversion of CL_{int} estimated from in vitro incubations to CL_{int} for the whole liver or body (Tables 7.2 and 7.3).

TABLE 7.2. Microsome and hepatocyte scaling factors for various species

	Human	Rat	Dog
MPPGL (mg microsomal protein per gram liver)	32 (95% CI: 29–34)	61 (95% CI: 47–75)	55 (95% CI: 48–62)
HPGL ($\times 10^6$ hepatocytes per gram of liver)	99 (95% CI: 74–131)	163 (95% CI: 127–199)	169 (95% CI: 131–207)

CI = confidence interval
Adapted from Smith et al. (2008) and Barter et al. (2007)

A detailed analysis of MPPGL and HPGL is not available for monkeys and, therefore, the human values of these parameters should be used for monkeys (Table 7.2).

TABLE 7.3. Liver weight and liver scaling factor (g liver per kg body weight) for various species

Species (weight)	Mouse (0.02 kg)	Rat (0.25 kg)	Dog (10 kg)	Monkey (5 kg)	Human (70 kg)
Liver weight (g)	1.75	10.0	320	150	1,800
Liver scaling factor (g liver per kg body weight)	87.5	40	32	30	25.7

Adapted from Davies and Morris (1993)

7.4.4 Liver Models of Hepatic Drug Clearance

Conversion of CL_{int} for the whole liver or body to $CL_{hepatic}$, which includes the impact of physiological factors such as blood flow and protein binding, involves the use of a liver model such as the well-stirred model, the parallel tube model, or the dispersion model.

None of the models has been shown to be superior to the others (Baranczewski et al. 2006); so, for simplicity, the equation for the well-stirred model is presented below:

$$CL_{\text{hepatic}} = Q \frac{f_u CL_{\text{int}}}{f_u CL_{\text{int}} + Q} \qquad (7.4)$$

CL_{hepatic} = hepatic clearance
f_u = unbound fraction
Q = hepatic blood flow

More details on the well-stirred model including values for hepatic blood flow are presented in Chap. 1.

In vitro–in vivo extrapolation is usually "validated" by determining if the in vitro ADME data correctly predict the clearance observed in preclinical studies. Although this is valuable, human clearance can involve pathways distinctly different from preclinical species, which can render this "validation" of limited value.

Microsomal Binding Considerations

Use of the well-stirred model for basic and neutral drugs or compounds often does not require the inclusion of f_u in the calculation. For basic and neutral drugs or compounds, f_u may cancel out with f_{umic} (unbound fraction in microsomes) (Obach 1999). The well-stirred model equation including f_{umic} is as follows:

$$CL_{\text{hepatic}} = Q \frac{f_u \dfrac{CL_{\text{int}}}{f_{\text{umic}}}}{f_u \dfrac{CL_{\text{int}}}{f_{\text{umic}}} + Q} \qquad (7.5)$$

In contrast to plasma protein binding, microsomal binding, f_{umic}, is generally not species dependent, provided the microsomal protein concentration is the same (Zhang et al. 2010).

Equations have been developed to estimate f_{umic} using drug lipophilicity (Hallifax and Houston 2006)

$$f_{\text{umic}} = \frac{1}{1 + C \bullet 10^{0.072 \bullet log\, P/D^2 + 0.067 \bullet log\, P/D - 1.126}}$$

C = microsomal protein concentration in mg/mL
$log\, P/D$ = log P if molecule is a base (pK$_a$ >7.4) and log D$_{7.4}$ if molecule is an acid or neutral compound (pK$_a$<7.4).

Continued

This equation has also been extended to include calculation of the unbound fraction in hepatocytes (Kilford et al. 2008).

7.4.5 Allometry

Allometry was initially used to establish an *empirical* relationship between the body surface area of an animal and its body weight.

$$S = a \times W^{0.67} \tag{7.6}$$

S = surface area
a = coefficient of allometric equation
W = body weight

Subsequently, allometry has been used to quantitatively relate a range of morphological parameters and biological functions to body weight, and the generic equation is:

$$Y = a \times W^b \tag{7.7}$$

Y = morphological parameter or biological function
b = exponent of allometric equation

Currently, allometry is widely used to predict human clearance based on preclinical clearance values.

$$\text{CL} = a \times W^b \text{ or } \log(\text{CL}) = \log(a) + b \times \log(W) \tag{7.8}$$

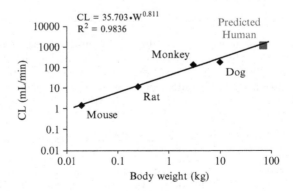

FIGURE 7.4. An example of allometry to predict human clearance.
CL = clearance.

An example is presented in Fig. 7.4. Note that the units for CL in this equation are mL/min, which subsequently needs to be converted to mL/min/kg. It is worth emphasizing that allometry is of little value if different clearance mechanisms are operational across species. Several types of correction factors have been proposed to improve the quality of these predictions. These factors decrease the predicted human clearance values relative to standard allometry.

Maximum life span (MLP) correction:

$$CL \times MLP = a \times W^b \tag{7.9}$$

Brain weight (BrW) correction:

$$CL \times BrW = a \times W^b \tag{7.10}$$

To clarify when to use a specific correction factor, the "rule of exponents" has been proposed as follows:

$$b < 0.7 \rightarrow \text{ standard allometry}$$

$$0.7 \leq b < 1.0 \rightarrow \text{ MLP correction}$$

$$b \geq 1.0 \rightarrow \text{ BrW correction}$$

A comprehensive analysis suggested that *"none of the correction factors resulted in substantially improved predictivity"* (Nagilla and Ward 2004).

Two species allometry has been proposed as well, and Tang et al. (2007) indicated that these methods were as predictive as three species allometry with the "rule of exponents."

$$CL_{human} = a_{rat-dog} \times W^{0.628} \tag{7.11}$$

$$CL_{human} = a_{rat-monkey} \times W^{0.650} \tag{7.12}$$

The unbound clearance (CL_u) can also be assessed in the allometric equation:

$$CL_u = a \times W^b \tag{7.13}$$

$$CL_u = CL/f_u$$

f_u = fraction unbound in plasma

Finally, Tang and Mayersohn (2005) proposed a method that enhances the predictivity of allometry, particularly in cases of "vertical allometry." Standard allometry is performed to determine the value of a (the coefficient of the allometric equation), and the human clearance is obtained using the following equation:

$$CL = 33.35 \times (a/Rf_u)^{0.77} \tag{7.14}$$

a = coefficient of standard allometric equation

$$Rf_u = f_{u,rat}/f_{u,human}$$

The following aspects should be taken into consideration when applying allometry:

- Allometry is empirical.
- Allometry is of little value if different clearance pathways are operational.
- Allometry may be most useful if the clearance is predominantly renally mediated across species.
- The R^2 value of the correlation should be considered when assigning a degree of confidence in the prediction.
- An exponent, significantly different from 0.7 (in particular <0.5 or >1.0) is indicative of significant species differences and should reduce the confidence in the prediction (Hu and Hayton 2001).
- The preclinical species with the lowest body weight (usually mouse or rat) and the highest body weight (usually dog) have the greatest influence on the human prediction.

Although some level of success has been achieved with allometry, some researchers have shown that the method is not more predictive than simpler approaches, such as single species scaling (Hosea et al. 2009) or the single species liver blood flow method (Nagilla and Ward 2004; Ward and Smith 2004).

7.4.6 Single Species Scaling

Single species scaling is based on direct extrapolation of the clearance in a particular preclinical species using a fixed exponent (usually 0.75), with or without a protein binding correction. In the former case, the equation is:

$$CL_{u,human} = CL_{u,animal} \times (W_{human}/W_{animal})^{0.75} \tag{7.15}$$

Opinions are divided about the most predictive preclinical animal model for humans. Hosea et al. (2009) observed good prediction of human pharmacokinetics using only rat data, which is also more practical than obtaining dog and/or monkey data. Other single species scaling methods were proposed by Tang et al. (2007) with the CL units being ml/min/kg.

$$CL_{human} = 0.152 \times CL_{rat} \tag{7.16}$$

$$CL_{human} = 0.410 \times CL_{dog} \tag{7.17}$$

$$CL_{human} = 0.407 \times CL_{monkey} \tag{7.18}$$

7.4.7 Single Species Liver Blood Flow Method

Clearance, as percentage of liver blood flow in preclinical species can be a predictor of human clearance. Ward and Smith (2004) suggested that the monkey is the most predictive species.

$$CL_{human} = CL_{animal} \times (Q_{human}/Q_{animal}) \tag{7.19}$$

7.5 HUMAN VOLUME OF DISTRIBUTION PREDICTION

V_d is predominantly governed by physicochemical parameters, and, consequently, species differences are less pronounced for V_d than for metabolic clearance. Therefore, human V_d is easier to predict than CL, and the differences between various methods are generally limited unless the V_d is remarkably small or large.

7.5.1 Allometry

Allometry has been used with some degree of success to predict human V_d.

$$V_d = a \times W^b \text{ or } \log(V_d) = \log(a) + b \times \log(W) \tag{7.20}$$

a = coefficient of allometric equation
b = exponent of allometric equation

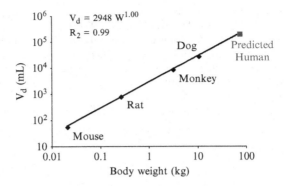

FIGURE 7.5. An example of allometry to predict human volume of distribution. V_d = volume of distribution.

An example is presented in Fig. 7.5. Note that the units for V_d in this equation are mL or L, which subsequently needs to be converted to mL/kg or L/kg. Correction factors, such as MLS and BrW, are generally not used. Some researchers have proposed that the unbound volume of distribution ($V_{d,u}$) should be used in the allometric equation as follows:

$$V_{d,u} = a \times W^b \tag{7.21}$$

$$V_{d,u} = V_d/f_u$$

f_u = fraction unbound in plasma

7.5.2 Single Species Scaling
Single species scaling is based on the similarity of V_d across species. Some have advocated using an identical V_d in humans as in preclinical species, but a correction factor is commonly applied to account for differences in plasma protein binding.

$$V_{d,u,human} = V_{d,u,animal} \times \left(f_{u,human}/f_{u,animal} \right) \tag{7.22}$$

7.5.3 The Oie–Tozer Method
In the Oie–Tozer method, the unbound fraction in human plasma and the average unbound fraction in tissues from preclinical species (assumed to be equal to the unbound fraction in human tissues), combined with appropriate human values for plasma and fluid volumes, are used to predict the human V_d (Obach et al. 1997).

$$V_{d,human} = V_p + (f_{u,human} \times V_e)$$
$$+ \left[(1 - f_{u,human}) \times \left(\frac{R_e}{i} \right) \times V_p \right] \quad (7.23)$$
$$+ \left(V_r \times \frac{f_{u,human}}{f_{ut,species\ average}} \right)$$

V_p = plasma volume
V_e = extracellular fluid volume
V_r = remainder of the fluid volume
R_e/i = ratio of binding proteins in extracellular fluid (except plasma) to binding proteins in plasma
$f_{ut,species\ average}$ = average unbound fraction in tissues from preclinical species

7.6 PHYSIOLOGICALLY BASED PHARMACOKINETIC MODELS

PBPK models offer a sophisticated integrated framework to understand compound disposition in the body and predict human pharmacokinetic profiles. The concept is described in detail in Chap. 10.

Prediction of Human Pharmacokinetic Profiles Using Methods Involving the Normalization of Animal Profiles

Methods predicting human pharmacokinetic profiles involving normalization of animal plasma concentration–time profiles include the Wajima plot and the various Dedrick methods. These methods are based on the assumption that plasma concentration–time profiles from different species are superimposable upon appropriate normalization of concentration and time scales. For illustrative purposes, a representative concentration–time profile and a Wajima plot are shown below. Each line represents the concentration–time profile of one animal species.

Continued

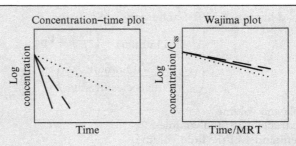

A Wajima plot involves normalization of the concentration scale by dividing it by the steady-state concentration (C_{ss}; calculated as Dose/V_{ss}]. The time scale is normalized by dividing it by the mean residence time (MRT; calculated as V_{ss}/CL). The composite profile of all preclinical species is used as the predicted human concentration–time profile. Ideally, normalized concentration–time profiles from various species superimpose upon each other, and predicted human profiles are obtained by back converting the normalized profile using predicted C_{ss} and MRT values for humans. The original set of compounds for which Wajima plots were applied were antibiotics with small V_d values that were mainly cleared renally in humans. A distinct advantage of this plot over the various Dedrick methods is that the CL and V_{ss} values used to calculate predicted C_{ss} and MRT values for humans can be obtained from any method of prediction. In contrast to the Wajima plot, the various Dedrick methods utilize allometric principles to normalize concentration and time scales. Detailed information on the Wajima plot and Dedrick methods can be found in Wajima et al. (2004) and Mahmood (2005).

7.7 CONTEXT OF CONFIDENCE IN HUMAN PHARMACOKINETIC PREDICTION

Human PK predictions are performed extensively preclinically and can determine the fate of compounds under consideration for development. At an earlier stage, predictions are based on limited data sets (in vitro and rodent in vivo data), and the goal is usually to categorize compounds and identify those that are worth pursuing further in preclinical studies. Once a more complete data set is available, it is possible to make more refined human PK predictions using perhaps more than one method, but even at this stage the predictions are more a reflection of assumed risk. Prospective predication of human pharmacokinetics continues to be an extremely difficult endeavor. For example, the most successful clearance methods generally have success rates of 60–80% with success defined as being within twofold of the observed value in humans. Often, having more preclinical data does not improve the prediction success rate. Indeed, Beaumont and Smith (2009) commented that *"Generation of further large amount of preclinical information on a compound with uncertain human pharmacokinetic prediction tends to add confusion rather than clarity."* Finally, the predicted PK properties of each drug candidate must be carefully evaluated in the context of its other properties and liabilities (i.e., in vivo potency, toxicity, etc.).

References

Akabane T, Tabata K, Kadono K et al (2010) A comparison of pharmacokinetics between humans and monkeys. Drug Metab Dispos 38:308–316

Baranczewski P, Stanczak A, Sundberg K et al (2006) Introduction to in vitro estimation of metabolic stability and drug interactions of new chemical entities in drug discovery and development. Pharmacol Rep 58:453–472

Barter ZE, Bayliss MK, Beaune PH et al (2007) Scaling factors for the extrapolation of in vivo metabolic drug clearance from in vitro data: reaching a consensus on values of human microsomal protein and hepatocellularity per gram of liver. Curr Drug Metab 8:33–45

Beaumont K, Smith DA (2009) Does human pharmacokinetic prediction add significant value to compound selection in drug discovery research? Curr Opin Drug Disc Dev 12:61–71

Chiou WL, Barve A (1998) Linear correlation of the fraction of oral dose absorbed of 64 drugs between humans and rats. Pharm Res 15:1792–1795

Chiou WL, Jeong HY, Chung SM et al (2000) Evaluation of using dog as an animal model to study the fraction of oral dose absorbed of 43 drugs in humans. Pharm Res 17:135–140

Chiou WL, Buehler PW (2002) Comparison of oral absorption and bioavailability of drugs between monkey and human. Pharm Res 19:868–874

Davies B, Morris T (1993) Physiological parameters in laboratory animals and humans. Pharm Res 10:1093–1095

Hallifax D, Houston JB (2006) Binding of drugs to hepatic microsomes: comment and assessment of current prediction methodology with recommendation for improvement. Drug Metab Dispos. 34:724–726

Hu T-M, Hayton WL (2001) Allometric scaling of xenobiotic clearance: uncertainty versus universality. AAPS PharmSci 3:1–14

Kilford PJ, Gertz M, Houston JB, Galetin A (2008) Hepatocellular binding of drugs: correction for unbound fraction in hepatocyte incubations using microsomal binding or drug lipophilicity data. Drug Metab Dispos. 36:1194–1197

Mahmood I (2005) Interspecies pharmacokinetic scaling: principles and application of allometric scaling. Pine House Publishers, Rockville, Maryland

Nagilla R, Ward KW (2004) A comprehensive analysis of the role of correction factors in the allometric predictivity of clearance from rat, dog, and monkey to humans. J Pharm Sci 93:2522–2534

Obach RS, Baxter JG, Liston TE et al (1997) The prediction of human pharmacokinetic parameters from preclinical and in vitro metabolism data. J Pharmacol Exp Ther 283:46–58

Obach RS (1999) Prediction of human clearance of twenty-nine drugs from hepatic microsomal intrinsic clearance data: an examination of in vitro half-life approach and nonspecific binding to microsomes. Drug Metab Dispos 27:1350–1359

Smith R, Jones RD, Ballard PG et al (2008) Determination of microsome and hepatocyte scaling factors for in vitro/in vivo extrapolation in the rat and dog. Xenobiotica 38:1386–1398

Tang H, Mayersohn M (2005) A novel method for prediction of human drug clearance by allometric scaling. Drug Metab Dispos 33:1297–1303

Tang H, Hussain A, Leal M et al (2007) Interspecies prediction of human drug clearance based on scaling data from one or two animal species. Drug Metab Dispos 35:1886–1893

Wajima T, Yano Y, Fukumura K et al (2004) Prediction of human pharmacokinetic profile in animal scale up based on normalizing time course profiles. J Pharm Sci 93:1890–1900

Ward KW, Smith BR (2004) A comprehensive quantitative and qualitative evaluation of extrapolation of intravenous pharmacokinetic parameters from rat, dog, and monkey to humans. I Clearance Drug Metab Dispos 32:603–611

Zhang Y, Yao L, Lin J et al (2010) Lack of appreciable species differences in nonspecific microsomal binding. J Pharm Sci 99:3620–3627

Additional Reading

De Buck SS, Mackie CE (2007) physiologically based approaches towards the prediction of pharmacokinetics: in vitro–in vivo extrapolation. Expert Opin Drug Metab Toxicol 3:865–878

Houston JB, Carlile DJ (1997) Prediction of hepatic clearance from microsomes, hepatocytes and liver slices. Drug Metab Rev 29:891–922

McGinnity DF, Collington J, Austin RP et al (2007) Evaluation of human pharmacokinetics, therapeutic dose and exposure predictions using marketed oral drugs. Curr Drug Metab 8:463–479

Obach RS (2001) The prediction of human clearance from hepatic microsomal metabolism data. Curr Opin Drug Discov Devel 4:36–44

Pelkonen O, Turpeinen M (2007) In vitro–in vivo extrapolation of hepatic clearance: biological tools, scaling factors, model assumptions and correct concentrations. Xenobiotica 37:1066–1089

Chapter 8
Advances in Bioanalysis as It Relates to ADME

Abstract
Bioanalysis continues to play a crucial role in ADME sciences. Most in vitro or in vivo studies have either a quantitative or qualitative bioanalytical endpoint. This chapter provides a general overview of the use of mass spectrometry in ADME sciences and the advantages and disadvantages of the various types of instruments.

Contents

8.1 ABBREVIATIONS

ADME	Absorption, distribution, metabolism, and excretion
APCI	Atmospheric pressure chemical ionization
API	Atmospheric pressure ionization
APPI	Atmospheric pressure photo ionization
DART	Direct analysis in real time
DBS	Dried blood spot
DESI	Desorption electrospray ionization
ESI	Electrospray ionization
GC	Gas chromatography
GLP	Good laboratory practice

S.C. Khojasteh et al., *Drug Metabolism and Pharmacokinetics Quick Guide*, DOI 10.1007/978-1-4419-5629-3_8,
© Springer Science+Business Media, LLC 2011

HILIC	Hydrophilic interaction liquid chromatography
HPLC	High performance liquid chromatography
LC	Liquid chromatography
MALDI	Matrix-assisted laser desorption/ionization
MIM	Multiple ion monitoring
MRM	Multiple reaction monitoring
MS	Mass spectroscopy
P450	Cytochrome P450
PD	Pharmacodynamic
PK	Pharmacokinetic
SIM	Selected ion monitoring
SRM	Selected reaction monitoring
UHPLC	Ultra high performance liquid chromatography or ultra high pressure liquid chromatography
UPLC	Ultra performance liquid chromatography
UV	Ultraviolet

8.2 BASIC CONCEPTS

Advances in bioanalytical sciences have played a critical role in improved integration of ADME (absorption, distribution, metabolism and excretion) sciences in drug discovery and development. Prior to the availability of LC–MS (liquid chromatography–mass spectrometry) techniques, most bioanalyses involved UV (ultra violet) analysis. The limited selectivity and sensitivity of UV analysis necessitated extensive sample clean up and long chromatographic gradients to increase selectivity. Therefore, most analyses were limited to determination of drug levels in support of toxicology and clinical studies. The introduction of LC–MS in the early 1990s allowed routine drug determination of drug levels in preclinical PK (pharmacokinetic) and efficacy studies as well in in vitro ADME studies.

Bioanalysis can be subdivided in to five sequential steps:

- Sample collection
- Sample extraction
- Chromatographic separation
- Bioanalysis by LC–MS
  - Ionization
  - One or more stages of mass analysis
  - Fragmentation (if MS/MS is employed)
  - Detection
- Data processing (not discussed here)

8.3 SAMPLE COLLECTION

For in vivo studies, blood is collected at specific timepoints. Reliable automated blood sampling devices are available for rodent studies. However, blood sampling for dog and monkey PK studies is still manual. Frequently, the blood is centrifuged to obtain plasma for analysis. In vitro ADME studies, such as P450 (cytochrome P450) inhibition or metabolic stability in microsomes or hepatocytes, are frequently performed using automated assays involving liquid handlers, and samples are collected at specific timepoints.

8.4 SAMPLE EXTRACTION

Because of the selectivity and sensitivity of LC–MS equipment, sample extraction from plasma and from samples of in vitro studies is usually limited to protein precipitation using 96-well or 384-well plates. If a low detection limit is required or if interference from an endogenous component occurs, more selective extraction procedures, such as liquid extraction and solid-phase extraction, may be required. Detailed steps of the sample extraction process are outlined in Table 8.1.

8.5 CHROMATOGRAPHIC SEPARATION

Reversed-phase high performance liquid chromatography (HPLC) is involved in most separation. The most common types of bonded phases are listed in Table 8.2. The analyte is retained by the bonded phase on the HPLC column and is eluted off the column using isocratic conditions or a gradient. With isocratic conditions, the % organic solvent and % water in the eluent is constant, which usually results in relatively broad signals due to limited chromatographic separation. When a gradient is employed, the % organic solvent is gradually increased over time to elute the analyte off the column, and chromatographic separation is improved. The availability of columns with particles less than 2 μm in diameter has improved chromatographic separation, but they do result in an increased column back pressure, and high pressure pumps are required. This technology is referred to as ultra performance liquid chromatography (UPLC) or ultra high performance/pressure liquid chromatography (UHPLC), and it offers significant advantages because it reduces the gradient time significantly without a loss in chromatographic separation (Plumb et al. 2008). An interesting alternative is the use of fused core particles. These particles have a solid core (about 1.7 μm

TABLE 8.1. Detail of plasma sample extraction via protein precipitation, liquid–liquid extraction and solid phase extraction

Step	Plasma protein precipitation	Liquid–liquid extraction	Solid phase extraction
1	Plasma	Plasma	Plasma
2	Add organic solvent (e.g., methanol or acetonitrile) containing the internal standard	Add internal standard	Add internal standard
3	Vortex	Add immiscible organic solvent (e.g., ethyl acetate)	Condition solid phase extraction cartridges or 96-well plate with solvents
4	Centrifuge	Vortex	Load plasma sample on top of the sorbent
5	Transfer supernatant	Transfer organic layer	Remove liquid using vacuum
6	Bioanalysis	Evaporate organic solvent	Wash sorbent with additional volumes of water or other appropriate solvent
7		Reconstitute residue in small volume of solvent compatible with bioanalysis	Elute analyte off the sorbent with a strong organic solvent
8		Vortex	Evaporate organic solvent
9		Bioanalysis	Reconstitute residue in small volume of solvent compatible with bioanalysis
10			Vortex
11			Bioanalysis

in diameter) surrounded by a layer of porous silica (about 0.5 μm thick). This technology offers similar chromatographic efficiency to UPLC, but does not require high pressure pumps.

TABLE 8.2.   Most commonly employed HPLC column bonded phase types

Phase type	Examples
Alkyl	C4, C8, C18, C30 chains
Aryl	Phenyl
Alkyl or aryl cyano	
Polar-embedded phases	C8/18 carbamate, C8/18 amide, C8/18 sulfonamide, other polar groups
Fluorinated phases	Fluoroalkyl, fluorophenyl

Hydrophilic interaction liquid chromatography (HILIC) is a special type of normal phase chromatography and is an alternative to regular reversed-phase chromatography for retention and separation of highly polar analytes. The stationary phase is quite polar, which fuels retention of the polar analytes. The analytes are eluted by gradually increasing the % water in the mobile phase, and the analytes elute in order of increasing hydrophilicity, which is the exact opposite of reversed-phase chromatography.

8.6   BIOANALYSIS BY LC–MS

The four integral steps to bioanalysis by mass spectrometry are:

- Ionization
- Mass analysis
- Fragmentation (if MS/MS is employed)
- Detection

8.6.1 Ionization

The first step in bioanalysis is evaporation of the solvent and ionization of the analytes. The two most common ionization techniques are electrospray ionization (ESI) and atmospheric pressure chemical ionization (APCI). Both ESI and APCI involve atmospheric pressure ionization (API).

8.6.1.1 Electrospray Ionization

In ESI, the HPLC column effluent elutes from a capillary carrying a high positive or negative voltage. This results in a Taylor cone as illustrated in Fig. 8.1 and the formation of small droplets with an excess positive or negative charge. A parallel and/or counter current of heated drying gas results in evaporation of solvent from the droplets, and the droplets become enriched in protonated ($[M + H]^+$ ions) or deprotonated ($[M - H]^-$ ions) analyte ions. The subsequent steps are still the topic of debate. One possibility is that ionized analyte molecules are expelled from the charged droplets because of Coulombic repulsion. Alternatively, the droplets fragment into smaller droplets because of Coulombic repulsion, and this process proceeds until a single ionized analyte molecule is left.

ESI allows ionization of biomolecules such as proteins and oligonucleotides, and the resultant ions carry multiple charges (e.g., $[M + nH]^{n+}$), which reduces the mass-to-charge ratio. Small molecules usually carry a single charge.

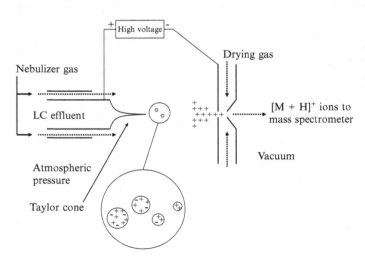

FIGURE 8.1. Electrospray ionization process in the positive ion mode.

8.6.1.2 Atmospheric Pressure Chemical Ionization
In APCI, the column effluent is very rapidly evaporated, and a discharge needle nearby generates a "cloud" of reagent ions formed from the solvent, which can transfer a positive or negative charge to the analyte of interest via proton transfer.

The question of whether to use APCI or ESI for quantitative bioanalysis has been debated at length, and some strengths and weaknesses of ESI and APCI are listed in Table 8.3. Nevertheless, both ionization techniques have been incorporated successfully in bioanalysis of small molecule drugs.

TABLE 8.3. Strengths and weaknesses of electrospray ionization and atmospheric pressure chemical ionization

	ESI	APCI
Advantages	Compatible with liquid chromatography	Compatible with liquid chromatography
	Ionization of small and large relatively polar molecules	Ionization of small non-polar or moderately polar molecules
Disadvantages	Prone to ion suppression effects, which reduces sensitivity and reproducibility	Thermal degradation of temperature-sensitive analytes
	Reduced ionization of moderately polar analytes	

In contrast to older ionization techniques, such as electron impact ionization, both ESI and APCI are compatible with liquid chromatography and are considered mild, i.e., very little fragmentation occurs in the ion source. Thus, the protonated $[M + H]^+$ ions or deprotonated $[M - H]^-$ and are transferred intact into the vacuum of the mass spectrometer for analysis. If the column effluent contains a significant amount of sodium or potassium, $[M + Na]^+$ and/or $[M + K]^+$ ions can be detected as well.

Other ionization techniques, such as atmospheric pressure photo ionization (APPI), desorption electrospray ionization (DESI), and direct analysis in real time (DART), are available as well, but not in common use. GC (gas chromatography)–MS combined with electron impact ionization is still preferred for analysis of volatile analytes.

8.6.2   Mass Analysis and Fragmentation

Mass spectrometers separate ions according to their mass-to-charge, m/z, ratio. Small molecules usually carry a single charge, and, therefore, their mass and mass-to-charge ratio are equivalent. Subsequent sections refer predominantly to the behavior of small molecules carrying a single charge. Multiple types of mass spectrometers are available, and each type has unique advantages and disadvantages.

8.6.2.1   MS Mode vs. MS/MS or MS^n Mode

Operation of a mass spectrometer in the MS mode allows separation of all ionized material according to mass. This provides a third dimension to the selectivity of the assay beyond sample extraction and chromatographic separation. However, the matrix can contain endogenous isobaric interference (i.e., ions with the same m/z ratio as the analyte). It is possible to address this by changing the sample extraction procedure or chromatographic separation, but it is easier to resolve this by switching to MS/MS analysis. The analyte of interest is selected with the first mass analyzer and fragmented by collision with an inert gas (e.g., helium, nitrogen, or argon). The second mass analyzer separates all fragment ions according to their mass-to-charge ratio. In the MS/MS mode, either a full scan spectrum can be acquired for metabolite identification or a specific fragment ion can be monitored for quantitative bioanalysis. The latter is referred to as selected reaction monitoring (SRM) or multiple reaction monitoring (MRM) if multiple transitions (e.g., analyte + internal standard or analyte + metabolite(s) + internal standard) are monitored. In SRM mode, it is much less likely that an endogenous interference has the same m/z value as the analyte as well as the same m/z value for the fragment ion that is monitored. This fourth dimension in selectivity has enabled much shorter cycle times (2 min or less per sample) and, therefore, increased throughput. With ion trap instruments it is possible to consecutively isolate and fragment ions, thereby allowing MS^n data to be generated. The advantages and disadvantages of MS and MS/MS are summarized in Table 8.4.

Thus, the selectivity of a reliable and sensitive quantitative LC–MS/MS assay is governed by the following parameters:

- Sample extraction
- Chromatographic separation
- First stage of mass selection
- Collision-induced fragmentation of ions and a second stage of mass selection

TABLE 8.4.   Advantages and disadvantages of analysis in the MS and MS/MS modes

	MS mode	MS/MS mode
Advantages	Ease of use	Increased selectivity and decreased likelihood of interference; a short chromatography column is usually sufficient
Disadvantages	Decreased selectivity and increased likelihood of interference; a long chromatography column may be required	Tuning of mass spectrometer requires more time and expertise
		The instrument is more expensive

8.6.2.2   Single Quadrupole Mass Spectrometers

In single quadrupole mass spectrometers, the ions travel between four rods that carry a combination of AC and DC voltages, which results in ions with a specific m/z ratio reaching the detector while others are deflected away. A complete mass spectrum can be obtained by scanning the quadrupole. Alternatively, a particular m/z ratio may be monitored, and this is called selected ion monitoring (SIM). It is possible to monitor multiple m/z ratios sequentially (i.e., multiple ion monitoring (MIM)) to detect one or more analytes and the internal standard.

8.6.2.3   Triple Quadrupole Mass Spectrometers

Triple quadrupole mass spectrometers are the most widely used instruments for quantitative bioanalysis. A triple quadrupole mass spectrometer is comprised of two sets of quadrupoles separated by a collision cell. The ions selected by the first quadrupole are fragmented into structured characteristic fragment ions, and these are separated in the second quadrupole according to their mass-to-charge ratio. Triple quadrupole mass spectrometers are also quite powerful for metabolite identification because of their ability to perform unique scan modes. The principles of constant neutral loss scanning and precursor ion mode scanning are illustrated in Fig. 8.2. The advantages and disadvantages of single and triple quadrupole mass spectrometers are summarized in Table 8.5.

Triple Quadrupole Scan Modes

	Ionization	Fragmentation		Detection
Product ion spectrum	ABC^+ $\rightarrow$	$A^+ + BC$ $\rightarrow$ $AB + C^+$ $\rightarrow$		A^+ C^+
	ABD^+ $\rightarrow$	$A^+ + BD$ $\rightarrow$ $AB + D^+$ $\rightarrow$		A^+ D^+
Precursor ion spectrum	ABC^+ $\rightarrow$	$A^+ + BC$ $\rightarrow$ $AB + C^+$		A^+
	ABD^+ $\rightarrow$	$A^+ + BD$ $\rightarrow$ $AB + D^+$		A^+
Constant neutral loss spectrum	ABC^+ $\rightarrow$	$A^+ + BC$ $AB + C^+$ $\rightarrow$		C^+
	ABD^+ $\rightarrow$	$A^+ + BD$ $AB + D^+$ $\rightarrow$		D^+

FIGURE 8.2. Product ion, precursor ion, and constant neutral loss MS/MS modes feasible with a triple quadrupole mass spectrometer.

TABLE 8.5. Advantages and disadvantages of analysis in single and triple quadrupole mass spectrometers

	Single quadrupole mass spectrometers	Triple quadrupole mass spectrometers
Advantages	Ease of use Very sensitive in SIM mode	Ease of use Very sensitive in SRM mode Increased selectivity Constant neutral loss and precursor ion scanning available for metabolite identification
Disadvantages	Unit mass resolution Limited full scan sensitivity Limited selectivity, which can affect sensitivity	Unit mass resolution Limited full scan sensitivity

8.6.2.4   Three-Dimensional and Linear Ion Trap Mass Spectrometers

Three-dimensional and linear ion traps operate by using a combination of AC and DC voltages to trap ions. A mass spectrum can be obtained by destabilizing their path inside the trap and ejecting them out of the trap toward the detector. The ability to consecutively isolate and fragment ions enables acquisition of

MS^n spectra. This makes these instruments quite useful for metabolite identification if the exact structure of fragment ions is in doubt. However, ion traps are less commonly used for quantitative bioanalysis because their inability to perform SIM and SRM makes them less sensitive than single and triple quadrupole mass spectrometers.

8.6.2.5   Time-of-Flight Mass Spectrometers

The operation of time-of-flight mass spectrometers is based on the principle that all ions receive the same amount of kinetic energy upon entering the mass spectrometer (provided they all carry the same number of charges), and their velocity is a function of their mass. Since the length of the flight tube is fixed, low mass ions have a high velocity and a short flight time, whereas high mass ions have a low velocity and a long flight time. The mass resolving power is such that it is possible to measure the exact mass of ions (within 5–10 ppm of the calculated mass) instead of the nominal mass (with unit mass resolution). Table 8.6 provides the exact mass and isotopic abundance of atoms commonly found in drug molecules. If the exact mass of an unknown entity is available, it is feasible to narrow down the number of possible molecular formulas, which makes this technique powerful for metabolite identification. Frequently, the time-of-flight section is preceded by a quadrupole and a collision cell to enable MS/MS capabilities.

TABLE 8.6. Exact mass and abundance of specific isotopes of atoms commonly found in drug molecules

Atom	Exact mass	Isotope abundance (%)
^{1}H	1.0078	99.985
^{12}C	12.0000	98.9
^{13}C	13.0034	1.1
^{14}N	14.0031	99.6
^{16}O	15.9949	99.8
^{19}F	18.9984	100
^{32}S	31.9721	95.0
^{33}S	32.9715	0.8
^{34}S	33.9679	4.2
^{35}Cl	34.9689	75.5
^{37}Cl	36.9659	24.5
^{79}Br	78.9183	50.5
^{81}Br	80.9163	49.5

8.6.2.6 Fourier Transform and Orbitrap Mass Spectrometers

Fourier transform mass spectrometers use a cryogenically cooled superconducting magnet to trap ions. The ions circulate in a cell in the bore of the magnet with a frequency that characterizes its mass. By monitoring the frequency of the complex circular motion it is possible to obtain the exact mass of the analyte (within 5 ppm of the calculated mass). The three-dimensional nature of the device allows for acquisition of MS^n spectra.

The orbitrap mass spectrometer was introduced about 5 years ago. An orbitrap is comprised of two circular electrodes with a small space in between. The electrostatic forces on the ions are balanced by the centrifugal forces on the ions resulting in a stable circular motion of the ions. Like the Fourier transform mass spectrometer, the orbitrap allows accurate mass measurements, but is easier to operate. Currently, the orbitrap is preceded by an ion trap, thus enabling acquisition of MS^n spectra. The advantages and disadvantages of three-dimensional, linear ion trap, time-of-flight, Fourier transform, and orbitrap mass spectrometers are summarized in Table 8.7.

TABLE 8.7. Advantages and disadvantages of analysis in three-dimensional, linear ion trap, time-of-flight, Fourier-transform, and orbitrap mass spectrometers

	Three-dimensional and linear ion trap mass spectrometers	Time-of-flight mass spectrometers	Fourier transform and orbitrap mass spectrometers
Advantages	Ease of use	Accurate mass determination (within 5–10 ppm)	Accurate mass determination (within 5 ppm)
	MS^n capabilities for metabolite identification	Increased sensitivity in full scan MS and MS/MS modes due to increased duty cycle	MS^n capabilities for metabolite identification
	Compact Relatively cheap		
Disadvantages	Low sensitivity for drug quantitation	Low sensitivity for drug quantitation	Low sensitivity for drug quantitation

(*continued*)

TABLE 8.7 (CONTINUED)

Three-dimensional and linear ion trap mass spectrometers	Time-of-flight mass spectrometers	Fourier transform and orbitrap mass spectrometers
(SIM and SRM not feasible)	(SIM and SRM not feasible)	(SIM and SRM not feasible)
Constant neutral loss and precursor ion scanning not possible for metabolite identification	Constant neutral loss and precursor ion scanning not possible for metabolite identification	Constant neutral loss and precursor ion scanning not possible for metabolite identification
Limited full scan sensitivity	Difficult to operate	Difficult to operate
Unit mass resolution		Very expensive

8.7 APPLICATIONS

8.7.1 Quantitative In Vitro ADME Studies

Quantitative bioanalysis is the cornerstone of high capacity acquisition of in vitro ADME data, and triple quadrupole mass spectrometers are most commonly used for this purpose. The following in vitro ADME assays have been integrated in the screening cascades for projects in early as well as late drug discovery:

- Metabolic stability in microsomes, S9, hepatocytes or recombinant P450 enzymes
- Competitive P450 inhibition
- Mechanism-based or time-dependent P450 inhibition
- Plasma protein binding or drug binding in other matrices such as microsomes or brain
- Blood to plasma partitioning
- Permeability in Caco-2 or MDCK cells or PAMPA
- Efflux in MDCK or LLCPK cells that overexpress transporters such as p-glycoprotein and BCRP

Most studies do not involve preparation of a standard curve to determine the absolute concentration. Quantitation is usually relative to a $t = 0$ min sample, a –NADPH sample, or another control sample. The order and capacity in which these assays are integrated in screening cascades is issue driven and, therefore, varies from project to project. However, metabolic stability and competitive P450 inhibition are frequently tier one assays and are preferably performed in parallel with potency assays. To increase the capacity of these in vitro ADME assays, the following techniques can be deployed:

- Column switching or multiplexing using multiple parallel chromatography columns connected to a single mass spectrometer (see Fig. 8.3) ensures that the column effluent is monitored only around the time the analyte elutes, which allows more samples to be analyzed per unit of time.

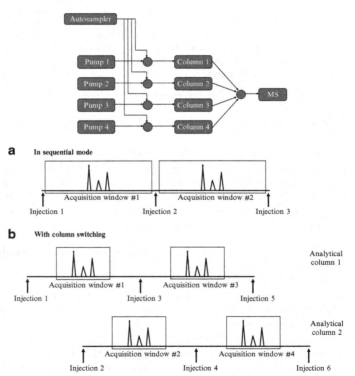

FIGURE 8.3. Column switching or multiplexing using multiple parallel chromatography columns connected to a single mass spectrometer. (**a**) Schematic of the connections between the autosampler, pumps, HPLC columns, and mass spectrometer for a system containing four parallel HPLC columns. (**b**) Data acquisition in the conventional sequential mode (**a**) and data acquisition using a system with two parallel HPLC columns (**b**).

- UPLC/UHPLC involves particles with a diameter smaller than 2 μm, and it increases chromatographic separation and/or reduces cycle time.
- Sample pooling combined with the use of MRM transitions to selectively monitor the various analytes of interest.

8.7.2 Quantitative In Vivo ADME Studies

LC–MS/MS is used extensively for bioanalysis of drugs in biological matrices such as plasma, blood, urine, and feces from in vivo studies. Triple quadrupole instruments are most powerful for this purpose, and they operate in the MRM mode monitoring the analyte, the internal standard and, if needed, metabolites. To obtain absolute drug levels, a standard curve is prepared and analyzed in the same batch of samples. Separately prepared quality control samples provide information about the ruggedness of the assay. Methods to support drug discovery studies can be developed in about 15 min, but methods to support GLP (good laboratory practice) toxicology and clinical bioanalysis require extensive validation (Vishwanathan et al. 2007; Savoie et al. 2010). To increase the capacity of in vivo pharmacokinetic screening, the following approaches can be deployed:

- *Sample pooling*: Samples from in vivo studies with different compounds are pooled and all MRM transitions are monitored.
- *Cassette or "n-in-one" dosing*: A mixture of up to five drugs is administered to the same animals and a triple quadrupole mass spectrometer operating in the MRM mode is used to quantitate all drugs. A risk associated with this technique is that one drug in the mixture may inhibit the metabolism of another drug. This risk can be reduced by reducing the dose. Frequently, a known reference compound is included in the cassette to gauge (albeit crudely) the extent of drug–drug interactions.
- *Column switching/multiplexing*: As illustrated in Fig. 8.3, column switching or multiplexing can be used to reduce the time spent on monitoring the effluent of a particular chromatography column.

The potential disadvantage of sample pooling and cassette dosing is that complex mixtures – containing multiple drug candidates and many metabolites – are analyzed. Thus, the risk of interference is more pronounced. This risk can be reduced to some extent by avoiding compounds with the same molecular weight or the same molecular weight as a likely metabolite of another compound (e.g., +16 Da or +32 Da metabolites).

> *Dried blood spot (DBS) analysis* has been used for a long time for sampling blood from newborns to detect metabolic disorders. Recently, DBS analysis has been combined with LC–MS/MS for quantitative bioanalysis (Spooner et al. 2009). This technique involves depositing a small amount of blood (<100 μl) on absorbent paper and allowing it to dry thoroughly. Subsequently, a small circle is punched out of the paper, which is transferred to a vial or 96-well plate for extraction with an organic solvent. The extract is analyzed by LC–MS/MS. The small amount of blood required for DBS analysis enables serial sampling of blood from mice, and it eliminates the need for a parallel PK group in rat toxicology studies. It also facilitates blood sampling from pediatric patients, and shipment of samples no longer requires dry ice.

8.7.3 Metabolite Identification

LC–MS/MS is a very powerful tool for metabolite identification. Before introduction of LC–MS/MS, it was hard to obtain structure-specific data for metabolites, and identification was usually based on similarity between the chromatographic retention time of the metabolite and an authentic standard. Although it may not be possible to identify the exact structure of the metabolite via LC–MS/MS, the information may be sufficient for chemists to address specific metabolic liabilities in the next generation of compounds. Later in the discovery stage, metabolite identification via LC–MS/MS can be used to identify the toxicology species with metabolic pathways most similar to humans. These studies are usually performed with liver microsomes, S9 or hepatocytes. It is also conceivable that a metabolite is responsible for part, or all, of the observed efficacy. Metabolic profiling may help resolve this PK–PD (pharmacokinetic–pharmacodynamic) disconnect. In the development stage, a greater understanding of the metabolic fate is desired and detailed LC–MS/MS studies are performed in particular to address concerns associated with the Metabolites in Safety Testing (MIST) guidance (see Chap. 6).

Interpretation of MS/MS spectra is still time-consuming. However, it is made easier by accurate mass measurements, which can help narrow down the number of possible molecular formulas for the metabolites and the assignment of fragment ions in MS/MS spectra (*vide supra*). Sometimes, it may be possible to propose the exact site of metabolism (e.g., *N*-dealkylation), but frequently identification is limited to a Markush structure in which the site of metabolic modification is associated with a part of the molecule. The latter is most common for hydroxylation reactions. To obtain unambiguous information it may be necessary to isolate the metabolite and obtain NMR data. Additional tools available to facilitate metabolite identification are:

- Chemical derivatization
- Hydrogen–deuterium exchange
- Studies with structurally related analogs

Finally, a metabolic profile obtained via LC–MS/MS is qualitative only. The ionization efficiencies can vary widely across metabolites, in particular if a basic center that enhances ionization in the positive ion mode has been eliminated from the drug candidate via metabolism (It is possible to use nanospray to reduce differences in ionization efficiency, but this technique is not routinely available (Hop et al. 2005)). Quantitative information can be obtained by splitting the flow and acquiring UV data (although this technique is not necessarily completely accurate either). Ideally, the drug is radiolabeled, and the flow is split to count absolute radioactivity.

Mass Defect Filtering
    All molecules have a specific mass defect, which reflects the difference between the exact mass and nominal mass of all atoms (see Table 8.6). For example, the exact mass of the $[M + H]^+$ ions of muraglitazar ($[C_{29}H_{29}N_2O_7]^+$) is 517.1967 Da and the mass defect is 0.1967 Da. Most phase 1 metabolites, in particular mono- or di-hydroxylated metabolites, have a mass defect fairly similar to that of the parent compound. Thus, full-scan high-resolution MS data can be filtered using a specific mass defect window close to that of the parent compound (Zhang et al. 2007, 2009). Note that phase 2 metabolites, such as sulfates or glucuronides, change the mass defect considerably, and, therefore, different filters need to be deployed to facilitate detection of these metabolites.

8.7.4 MALDI Tissue Imaging

Knowing the distribution of drugs in tissues can facilitate explanation of observed efficacy or toxicity. Prior to a human mass balance study, a quantitative whole body autoradiography study is performed in rodents (and in a non-rodent if this animal species more closely resembles human pharmacokinetics) to see if drug-related material is retained in specific tissues or organs. This study requires radiolabeled material, and it is not possible to distinguish the parent compound from metabolites. The latter two disadvantages can be circumvented with matrix-assisted laser desorption/ionization imaging (MALDI; Cornett et al. 2008; Khatib-Shahidi et al. 2006). With this technique, a solution containing a UV-absorbent matrix is sprayed on a tissue slice. After evaporation of the solvent, the tissue is transferred to the vacuum of the mass spectrometer and a laser is scanned across the tissue slices in discrete steps. This mass selective detection technique allows detection of both the parent compound and metabolites with a spatial resolution of up to 30 μm.

References

Cornett DS, Frappier SL, Caprioli RM (2008) MALDI-FTICR imaging mass spectrometry of drugs and metabolites in tissue. Anal Chem 80:5648–5653

Hop CECA, Chen Y, Yu LJ (2005) Uniformity of ionization response of structurally diverse analytes using a chip-based nanoelectrospray ionization source. Rapid Commun Mass Spectrom 19:3139–3142

Khatib-Shahidi S, Andersson M, Herman JL et al (2006) Direct molecular analysis of whole-body animal tissue sections by imaging MALDI mass spectrometry. Anal Chem 78:6448–6456

Plumb RS, Potts WB III, Rainville PD (2008) Addressing the analytical throughput challenges in ADME screening using ultra-performance liquid chromatography/tandem mass spectrometry methodologies. Rapid Commun Mass Spectrom 22:2139–2152

Savoie N, Garofolo F, Van Amsterdam P et al (2010) 2009 white paper on recent issues in regulated bioanalysis from the 3rd calibration and validation group workshop. Bioanalysis 2:53–68

Spooner N, Lad R, Barfield M (2009) Dried blood spots as a sample collection technique for the determination of pharmacokinetics in clinical studies: considerations for the validation of a quantitative bioanalytical method. Anal Chem 81:1557–1563

Vishwanathan CT, Bansal S, Booth B et al (2007) Quantitative bioanalytical methods validation and implementation: best practices for chromatographic and ligand binding assays. Pharm Res 24:1962–1973

Zhang D, Cheng PT, Zhang H (2007) Mass defect filtering on high resolution LC/MS data as a methodology for detecting metabolites with unpredictable structures: identification of oxazole-ring opened metabolites of muraglitazar. Drug Metab Lett 1:287–292

Zhang H, Zhang D, Ray K et al (2009) Mass defect filter technique and its application to drug metabolite identification by high resolution mass spectrometry. J Mass Spectrom 44:999–1016

Additional Readings

Chowdhury SK (2005) Identification and quantification of drugs, metabolites and metabolizing enzymes by LC-MS. Elsevier, Amsterdam, The Netherlands

Hop CECA (2006) LC-MS in drug disposition and metabolism. In: Caprioli RM (ed) The encyclopedia of mass spectrometry, vol 3. The Netherlands, Elsevier, Amsterdam, pp 233–274

Korfmacher WA (2010) Using mass spectrometry for drug metabolism studies, 2nd edn. CRC Press, Boca Raton, FL

Ramanathan R (2009) Mass spectrometry in drug metabolism and pharmacokinetics. Wiley, New York

Chapter 9
ADME Properties and Their Dependence on Physicochemical Properties

Abstract

Absorption, distribution, metabolism, and excretion (ADME) properties such as absorption, clearance, and volume of distribution are strongly influenced by physicochemical parameters. A lot of retrospective analyses have been performed to identify those attributes that give rise to favorable ADME parameters. These attributes should be incorporated in compound design to increase the chance of identifying a compound with superior ADME properties.

Contents

9.1   ABBREVIATIONS

ADME   Absorption, distribution, metabolism, and excretion
BCS   Biopharmaceutics classification system
CNS   Central nervous system

S.C. Khojasteh et al., *Drug Metabolism and Pharmacokinetics Quick Guide*, DOI 10.1007/978-1-4419-5629-3_9, © Springer Science+Business Media, LLC 2011

FaSSIF   Fasted state simulated intestinal fluid
FeSSIF   Fed state simulated intestinal fluid
HBA   Hydrogen bond acceptor
HBD   Hydrogen bond donor
iv   Intravenous
MV   Molecular volume
MW   Molecular weight
po   Oral
PSA   Polar surface area
RB   Rotatable bond
SA   Surface area
TPSA   Topological polar surface area

9.2 BASIC CONCEPTS

In vivo pharmacokinetic parameters, such as absorption, distribution, metabolism, and excretion are strongly influenced by the physicochemical properties of a drug. The earliest thorough analysis of ADME properties was performed by Lipinski and resulted in the famous "rule of 5", which argues that poor absorption is more likely if:

- The molecular weight (MW) >500 Da
- The number of hydrogen bond donors (HBDs) >5 (counting the sum of all NH and OH groups)
- Log P >5
- The number of hydrogen bond acceptors (HBAs) >10 (counting all N and O atoms)

Many analyses have been performed looking at the impact of physicochemical parameters on drug candidates. Although some companies do not pursue compounds that violate one or more components of the "rule of 5", the goal of these guidelines is not necessarily to rule out certain synthetic ideas. After all, several successful drugs violate the "rule of 5" to some extent: for example, atorvastatin, montelukast, and natural products such as cyclosporine and paclitaxel. These guidelines are more intended to steer the synthetic chemistry effort toward chemical space that is *more likely* to yield drugs with *superior* ADME properties. A brief summary of the physicochemical properties of marketed oral drugs relative to Lipinski's "rule of 5" is presented in Table 9.1.

TABLE 9.1. Average physicochemical properties of oral drugs in phase I and marketed oral drugs in comparison with Lipinski's "rule of 5"

	Phase I oral drugs	Marketed oral drugs	Oral drugs launched pre-1983	Oral drugs launched 1983–2002	Lipinski's "rule of 5"
MW (Da)	423	337	331	377	≤500
clog P	2.6	2.5	2.3	2.5	≤5
clog $D_{7.4}$	1.3	1.0			
Number of HBDs	2.5	2.1	1.8	1.8	≤5
Number of HBAs	6.4	4.9	3.0	3.7	≤10
Number of RBs	7.8	5.9	5.0	6.4	
Number of rings			2.6	2.9	

Data from Wenlock et al. (2003) and Leeson and Davis (2004).
clog P = calculated lipophilicity when the compound is neutral; clog $D_{7.4}$ = calculated lipophilicity at a pH of 7.4; RB = rotatable bond

Table 9.1 also highlights the fact that the average MW of oral drugs steadily decreases going from phase I to the market. This trend is also reflected in the number of HBDs and HBAs. (The same trend also applies when going from hit to lead to candidate in drug discovery.) However, the percentage of launched oral drugs that has a MW <350 Da has steadily decreased from 60–70% around 1985 to 30–40% around 2005 (Leeson and Springthorpe 2007), and there is a statistically significant difference between MW and the number of HBAs and rotatable bonds (RBs) between drugs launched before 1983 and between 1983 and 2002 (see Table 9.1). Note that nonoral drugs have different physicochemical properties. For example, injectable drugs have a higher number of HBDs, HBAs, and RBs and a higher average MW, as well as a lower mean clog P than oral drugs (Vieth et al. 2004).

Of course, ADME properties have to be balanced with other properties such as potency, selectivity, toxicity, etc. Indeed, significant differences exist in the average physicochemical properties of drugs as a function of the type of target such as ion channels, G protein-coupled receptors, proteases, kinases, etc. (Morphy 2006; Paolini et al. 2006). For some targets, such as those based on the disruption of protein–protein interaction, it is particularly unlikely to stay within the "rule of 5". In addition, the nature of the blood–brain barrier is such that cutoffs of certain physicochemical parameters, such as MW, number of HBAs, and RBs, and topological polar surface area (TPSA) need to be reduced for central nervous system (CNS) drugs. Physicochemical properties for oral CNS drugs are illustrated in Table 9.2.

TABLE 9.2. Average physicochemical properties of oral CNS drugs in phase I and marketed drugs in comparison with Lipinski's and Pajouhseh's rules for CNS compounds

	Oral CNS drugs launched 1983–2002	Oral CNS drugs launched 1983–2002	Lipinski's rule for CNS compounds	Pajouhseh's rule for CNS compounds
MW (Da)	310	377	≤400	<450
clog P	2.5	2.5	≤5	<5
clog $D_{7.4}$				
Number of HBDs	1.5	1.8	≤3	<3
Number of HBAs	2.1	3.7	≤7	<7
Number of RBs	4.7	6.4		<8
Number of rings	2.9	2.9		
PSA (Å^2)				60–90
% PSA	16	21		
pK_a				7.5–10.5

Data from Leeson and Davis (2004) and Pajouhesh and Lenz (2005)
PSA = polar surface area

It is important to keep in mind that the conclusions about correlations between physicochemical and ADME properties can be strongly influenced by the size and nature of the database employed. Moreover, *many of the parameters are not independent of each other*. Correlations between various parameters are illustrated in Table 9.3. For example, if MW increases, the number of RBs usually increases as well, and the TPSA and the total number of N and O atoms are also correlated (Vieth et al. 2004). Thus, it is possible that an observed correlation is merely fortuitous because other, more critical, properties have changed, which may actually be driving the correlation. Finally, some parameters may influence a range of ADME properties, while some may impact only one parameter. For example, the number of HBDs and HBAs has a pronounced impact on absorption but has a small impact on intestinal and hepatic extraction (Varma et al. 2010).

TABLE 9.3. Correlation coefficients between various physicochemical parameters

	MW	clog P	ON	OHNH	Number of atoms	Number of rings	Number of RBs	Total SA	PSA	HBAs*	HBDs*
MW		0.18	0.45	0.12	0.96	0.51	0.50	0.88	0.33	0.39	0.13
clog P	-0.03		-0.55	-0.40	0.23	0.20	0.09	0.33	-0.60	-0.51	-0.38
ON	0.82	-0.44		0.43	0.41	0.04	0.36	0.28	0.93	0.79	0.42
OHNH	0.66	-0.44	0.78		0.11	-0.07	0.12	0.06	0.54	0.34	0.99
Number of atoms	0.97	0.01	0.82	0.65		0.59	0.49	0.92	0.28	0.32	0.12
Number of rings	0.55	0.20	0.34	0.21	0.62		-0.29	0.38	-0.06	0.07	-0.05
Number of RBs	0.77	-0.10	0.72	0.62	0.77	0.16		0.70	0.25	0.17	0.11
Total SA	0.96	0.05	0.78	0.64	0.98	0.54	0.84		0.14	0.18	0.07
PSA	0.74	-0.53	0.96	0.82	0.72	0.24	0.67	0.68		0.81	0.53
HBAs*	0.70	-0.46	0.87	0.64	0.67	0.26	0.54	0.62	0.88		0.32
HBDs*	0.66	-0.42	0.77	1.00	0.66	0.22	0.62	0.64	0.81	0.62	

Correlation coefficients (r) for a dataset of 1,719 marketed drugs are displayed in the lower diagonal and r values for a subset that satisfies 10–90% MW coverage (196–563 Da) are displayed in the upper diagonal.
Data from Vieth et al. (2004)
ON = number of oxygen and nitrogen atoms; OHNH = number of OH and NH groups; total SA = total surface area
*HBA is defined slightly different than ON, and HBD is defined slightly different than OHNH. See Vieth et al. (2004) for details.

The list of physicochemical parameters that influence ADME properties includes MW, pK_a, log P, log $D_{7.4}$, TPSA, % TPSA, HBAs, HBDs, (aromatic) rings, % sp^3 carbon atoms, RBs, and solubility and we will focus on the correlation between these physicochemical parameters and their ADME attributes. A detailed explanation of the physicochemical underpinning of ADME properties is beyond the scope of this book.

9.3 MOLECULAR WEIGHT

MW can be calculated easily and is quite relevant from an ADME point of view. Elaborate studies have shown that permeability decreases with increasing MW; this observation fueled Lipinski to propose a cut off of 500 Da for potential drugs. However, some molecules with a MW >500 Da are absorbed. Many natural products have a MW >500 Da, but there are indications that absorption of some of these compounds may be mediated by uptake transporters. In some cases, MW is deliberately increased (and may exceed 500 Da) by making a prodrug to improve permeability. For example, olmesartan medoxomil is an ester prodrug and is absorbed well, whereas the active, dianionic drug is poorly absorbed (Fig. 9.1).

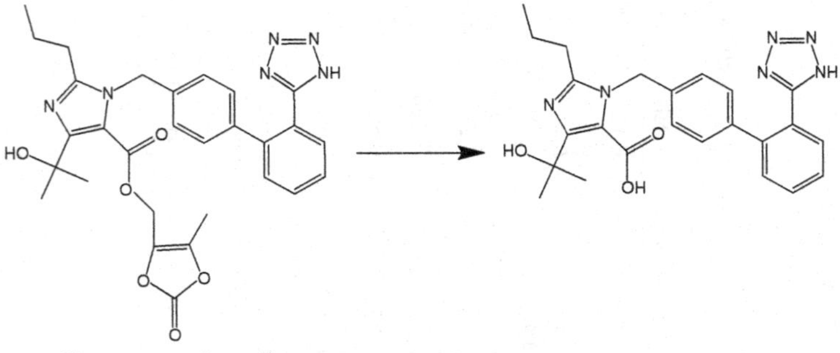

Olmesartan medoxomil

FIGURE 9.1. Conversion of the prodrug olmesartan to the active drug.

MW is also loosely correlated with clearance whereby clearance increases with increasing MW. The reason may simply be that the number of metabolically reactive regions in molecules (also called "soft spots") increases with MW.

It has been argued that molecular volume (MV) is more relevant for the rate and extent of oral absorption and tissue distribution than MW (Lobell et al. 2006). MV is related to MW and can be obtained using (9.1).

$$MV = MW/1.336 \tag{9.1}$$

However, halogen atoms should be accounted for differently because they have a relatively small volume for their atomic mass. To account for this effect, the following *corrected* atomic weights should be used for fluorine, chlorine, bromine, and iodine when calculating the corrected MW of halogen containing drugs: 5.2, 19.2, 26.3, and 37.4 Da (Lobell et al. 2006). (The real atomic weights of fluorine, chlorine, bromine, and iodine are 19.0, 35.5, 79.9, and 126.9 Da.) Many drugs contain multiple halogen atoms, such as fluorine and chlorine, to improve metabolic stability and/or the interaction with the target, but these drugs may have a MW >500 Da and, hence, violate Lipinski's "rule of 5". However, if the corrected atomic weights are used for the halogen atoms, the corrected MW may be reduced significantly and explain the favorable ADME properties. For example, amiodarone contains two iodine atoms and has a MW of 645 Da. However, the corrected molecular weight is 466 Da. (In addition, amiodarone has a log P of 8.9, but the log $D_{7.4}$ is much lower at 3.4.) Indeed, the ADME properties of this compound are good: the bioavailability is 30% and the clearance is about 2 ml/min/kg (Chow 1996; Doggrell 2001) (Fig. 9.2).

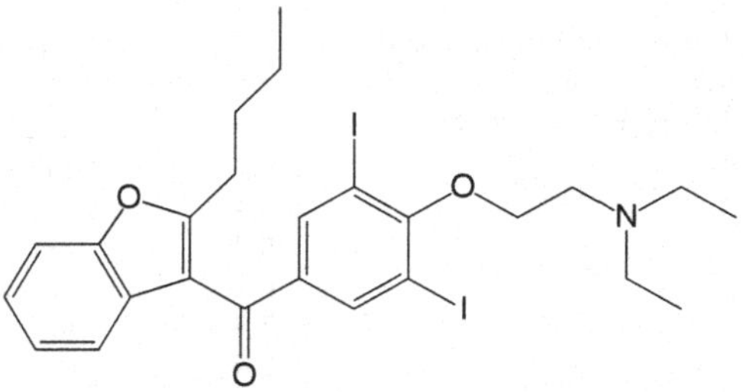

FIGURE 9.2. Structure of amiodarone.

Amiodarone

9.4 pK_a

The degree of ionization of the drug is determined by the pH of the medium (acidic in the stomach and upper intestine, but close to 7.4 systemically) and the pK_a.

For acids:

$$AH + H_2O \leftrightarrow A^- + H_3O^+ \tag{9.2}$$

$$pK_a = -\log([A^-][H_3O^+]/[AH]) \qquad (9.3)$$

For bases:

$$BH^+ + H_2O \leftrightarrow B + H_3O^+ \qquad (9.4)$$

$$pK_a = -\log([B][H_3O^+]/[BH^+]) \qquad (9.5)$$

Knowing that pH $= -\log[H_3O]^+$, (9.3) and (9.5) can be converted as shown below.
For acids:

$$pH = pK_a + \log[A^-] - \log[AH] \qquad (9.6)$$

For bases:

$$pH = pK_a + \log[B] - \log[BH^+] \qquad (9.7)$$

These equations indicate that, generally, acids (AH) will be neutral in the stomach, but the conjugated base (A$^-$) will predominate in the intestine and blood. For bases, the acidic form (BH$^+$) will predominate in the stomach, but basic compounds will be neutral in the intestine and blood, unless the compound has a pK_a substantially greater than 7.4, as illustrated in Table 9.4.

TABLE 9.4. Ionization at pH = 7.4 for acids and bases as a function of pK_a

	Ionization at pH 7.4		
	Acid		Base
pK_a	% ionized	pK_a	% ionized
4.4	99.9	5.4	1
5.4	99	6.4	10
6.4	90	7.4	50
7.4	50	8.4	90
8.4	10	9.4	99
9.4	1	10.4	99.9

The pK_a value strongly influences solubility and drug absorption and distribution because it is assumed that only the neutral species can cross the lipophilic membrane. This phenomenon is illustrated in Fig. 9.3. In keeping with this phenomenon, acids tend to have a low volume of distribution (although this is also influenced by plasma protein binding) because the anionic state predominates in the plasma. In contrast, compounds with basic moieties, such as amines, tend to have a high volume of distribution. This is to a significant extent fueled by (1) sequestration in membranes (due to interaction with anionic phospholipids) or (2) trapping in acid organelles such as lysozomes.

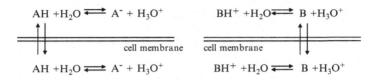

FIGURE 9.3. Impact of ionization state on membrane permeability for an acid and a base.

9.5 LIPOPHILICITY

Lipophilicity reflects the affinity of a drug molecule to be associated with a nonpolar lipid-rich medium (in contrast to preferentially residing in an aqueous medium). Lipophilicity is measured by determining the partitioning between a buffered aqueous phase and an organic phase, usually n-octanol, and is expressed as either log P or log D. Log P reflects the partitioning when the compound is neutral (i.e., not charged), whereas log D is measured at a specific pH at which a fraction of the compound may be neutral and the rest positively or negatively charged depending on the pH and the pK_a.

$$\log P = \log([compound_{organic\ phase}]/[compound_{aqueous\ phase}]) \quad (9.8)$$

$$\log D_{pH} = \log([compound_{organic\ phase}]/[compound_{aqueous\ phase}]) \quad (9.9)$$

Log P is included in Lipinski's "rule of 5" because at high log P values (>5), solubility is generally poor, which prevents absorption, and/or molecules partition into membranes and will not cross the enterocytes. Log P can be calculated easily (log P reflects the calculated value) and relatively accurately (usually ±1), but calculation of log D_{pH} is more complicated, because it requires knowledge of the pK_a. This explains why emphasis has been placed on determining log P. However, from a physiological perspective, log D_{pH} is more relevant. For example, ebastine, a nonsedating H1 antihistamine, has a high log P of 6.9, but its log $D_{7.4}$ is much lower at 4.6, because it is a basic amine and ebastine is partially charged at neutral pH. Log D_{pH} is usually determined at pH $= 6.0$–7.4, which reflects the pH in the intestine (Fig. 9.4).

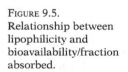

Ebastine

FIGURE 9.4. Structure of ebastine.

There appears to be a Gaussian or parabolic relationship between $\log D_{7.4}$ and the extent of absorption and bioavailability (see Fig. 9.5). At $\log D_{7.4} < 0$, a compound's solubility is good, but its permeability across membranes is poor, resulting in limited absorption. In addition, the metabolic clearance is usually limited, but clearance by the kidney may be high. In contrast, at $\log D_{7.4} > 5$, membrane permeability is adequate, but the solubility is low, which significantly reduces absorption. Moreover, metabolic clearance tends to be higher for compounds with $\log D_{7.4} > 5$. This effect is exacerbated by increased plasma protein binding (i.e., the unbound clearance is much greater at high $\log D_{7.4}$). However, this effect appears to be modulated by MW such that the $\log D$ range that leads to good absorption and increased metabolic stability is wider at low MW than at high MW (Johnson et al. 2009).

FIGURE 9.5.
Relationship between
lipophilicity and
bioavailability/fraction
absorbed.

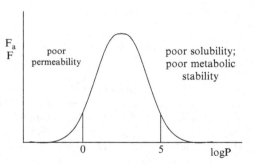

Plasma protein binding and tissue binding increases with increasing $\log P$ and $\log D_{7.4}$, and this can lead to a relatively low free drug concentration at the site of action despite reasonable membrane permeability.

Finally, there is a correlation between toxicity and clog P and TPSA with the observed odds for toxicity dramatically increasing with clog P >3 and TPSA <75 $Å^2$ (Price et al. 2009). This effect has been ascribed to increased off-target activity for nonpolar lipophilic compounds (Leeson and Springthorpe 2007; Price et al. 2009).

9.6 TOPOLOGICAL POLAR SURFACE AREA

Calculating the exact polar surface area (*3D PSA*) can be time consuming because it involves calculating the three-dimensional structure and the PSA itself. The easiest and fastest approach is calculating the TPSA, which involves summing the contributions of individual polar fragments (Ertl et al. 2000). Although TPSA is based solely on the two-dimensional structure, the correlation between 3D PSA and TPSA was shown to be 0.99 for 34,810 molecules from the World Drug Index.

As described above, TPSA is an important parameter for passive membrane permeability, with a TPSA in excess of 120 $Å^2$ associated with poor absorption and a value in excess of 90 $Å^2$ associated with poor brain penetration. This effect can be ascribed to decreased membrane permeability for polar compounds. In contrast, a very low TPSA (<50 $Å^2$) can lead to increased intestinal and hepatic extraction (Varma et al. 2010), increased risk of off-target activity, and increased toxicity (vide supra; Price et al. 2009).

9.7 NUMBER OF HYDROGEN BOND DONORS AND ACCEPTORS

For the sake of ease of calculation, the number of HBDs and HBAs are usually equated with the sum of all NH and OH groups and the sum of all N and O atoms, respectively. The number of HBDs and HBAs is clearly correlated with the TPSA, and it has been established that the permeability and, hence, absorption, reduces with an increase in the number of HBDs and HBAs. On the basis of a database of 309 compounds with iv and po data in humans, Varma et al. (2010) showed that the fraction absorbed reduces significantly once the number of HBDs and HBAs exceeds 10, whereas the intestinal and hepatic extraction slightly decreases with increasing number of HBDs and HBAs. For CNS drugs, it is important to keep in mind that P-glycoprotein efflux increases dramatically with increasing number of HBDs (see Table 9.2).

9.8   NUMBER OF (AROMATIC) RINGS AND % SP^3 CARBON ATOMS

The number of (aromatic) rings has a significant correlation with solubility, log P and plasma protein binding, and these parameters influence ADME attributes. For example, based on the analysis of a proprietary compound collection at GlaxoSmithKline, it was shown that 80% of the compounds with two aromatic rings have a clog P <3, whereas only 17% of the compounds with five aromatic rings have a clog P <3 (Ritchie and Macdonald 2009). The authors argued that compounds with more than three aromatic rings were associated with an increased risk of compound attrition and, therefore, should be avoided. However, the effect seems more pronounced for carboaromatic than metero aromatic rings (Ritchie et al. 2011) Similarly, others have shown a strong correlation between the fraction of sp^3 atoms and solubility and melting point (Lovering et al. 2009). In addition, they showed that the % sp^3 atoms steadily increases going from discovery to marketed drugs (36% versus 47%).

9.9   SOLUBILITY

Solubility is obviously of great importance for absorption of oral drugs. However, this parameter still remains somewhat elusive despite its apparent simplicity. Several aspects should be considered.

1. There are two types of solubility measurements: kinetic and thermodynamic solubility. The kinetic solubility is obtained by dissolving the compound in an organic solvent (e.g., DMSO) and adding it to aqueous buffer. Equilibrium is not reached between the dissolved compound and the solid, which may not be the most stable polymorph. Kinetic solubility measurements may be useful to assess the lack of solubility encountered in routine in vitro potency and ADME assays. However, kinetic solubility measurements are of little relevance to the situation encountered in vivo. Thermodynamic solubility is obtained by adding the aqueous buffer directly to solid crystalline material and waiting for an extended period of time for equilibrium between the dissolved and solid material. Although thermodynamic solubility is more relevant, it consumes more material and the measurement is time consuming.

2. The first synthetic batch is frequently amorphous, which usually results in increased solubility. Even if subsequent batches are crystalline, extensive experimentation may be required to identify the most stable polymorph, and the most stable

polymorph is usually not identified until the compound has been nominated for development. Frequently, the most stable polymorph will have a much higher melting point and reduced solubility.

3. Solubility is pH dependent. For example, moderately basic compounds may have very good solubility at pH = 1–2 (i.e., the situation encountered in the stomach), but the solubility may be much lower at the pH encountered in the intestine (see Fig. 9.6).

4. Solubility also depends on the matrix, and, therefore, thermodynamic solubility is usually measured in buffer, fasted state simulated intestinal fluid (FaSSIF; $pH \approx 6.5$), and fed state simulated intestinal fluid (FeSSIF; pH $\approx$ 5). For example, felodipine is 100-fold more soluble in FeSSIF than in water and is not pH dependent in water.

5. It is possible to improve the dissolution of a drug by changing the salt form, reducing the particle size, or modifying the formulation (e.g., adding excipients to improve the solubility of lipophilic drugs). This enhances the solubility in the stomach and hopefully the compound stays in solution while it enters and moves along the intestine.

6. Not only is the extent of solubility important, but the rate should also be taken into consideration. The dissolution rate can be assessed in thermodynamic solubility experiments.

7. Solubility and dissolution rate are influenced by several physicochemical parameters listed above such as pK_a, lipophilicity, etc.

8. A good computational model to calculate solubility reliably is still not available because solubility depends heavily on the crystal structure. Although it is possible to predict the crystal structure of small organic molecules (Day et al. 2009), such a prediction requires a massive amount of computing and is currently used only to predict polymorphs of compounds in development.

Poor solubility negatively affects absorption, in particular if permeability is poor to moderate. A key parameter to consider in this context is the size of the dose. In the biopharmaceutics classification system (BCS; see Chap. 3) a compound is considered highly soluble if the dose can be dissolved in 250 ml. The situation encountered in toxicology studies should be taken into consideration as well. It is possible that the solubility is compatible with absorption of the human dose, but it may not be possible to

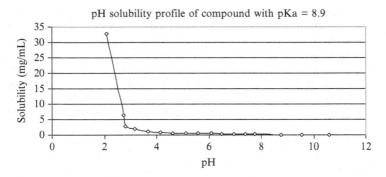

FIGURE 9.6. Solubility profile of a basic compound with $pK_a = 8.9$.

increase the exposure sufficiently to get the desired therapeutic index because of poor solubility.

9.10   MULTIPARAMETER OPTIMIZATION

Instead of looking at individual parameters, it is more advantageous to look at multiple parameters and, preferably, to weigh each parameter. This approach can be based on physicochemical parameters that can be calculated *prior to synthesis*, but it is also useful to incorporate *measured* parameters (such as enzyme and cell potency, metabolic stability in microsomes and hepatocytes, P450 inhibition, etc.) to facilitate identification of lead compounds and series in an objective fashion. A simple scoring system was proposed by scientists at Bayer and is depicted in Table 9.5 (Lobell et al. 2006). The final score is obtained by summation of the scores for each individual parameter with the best score being 0 and the worst score being 10. The authors showed that 70% of a database of 812 oral drugs had a score of 2 or less, whereas the average score of 13,775 confirmed high-throughput screening hits was 4.1. A more sophisticated approach has been advocated by Segall et al. (2009) who incorporated probabilistic scoring that allows parameters in the scoring profile to be given a different importance and it considers the uncertainty in each data point which is critical to distinguish molecules with confidence.

Although multiparameter optimization is empirical, two areas of controversy that continuously arise are (1) the weighing of the individual parameters and (2) the procedure applied to combine the parameters. The simplest and most common procedures for the latter are summation and multiplication. It is possible that a compound has reasonable scores in most areas, but a poor score in one particular area. If the individual scores are summed, a reasonable final score will be obtained for this compound despite the poor score in one possibly critical area. However, the final score will be small if the individual scores are multiplied.

TABLE 9.5. Simple scoring algorithm to categorize hits and leads

Score	Solubility (mg/L)	clog P	MW$_{corrected}$ (Da)	PSA (Å^2)	Number of RBs
0	≥50	≤3	≤400	≤120	≤7
1	10–50	3–5	400–500	120–140	8–10
2	<10	>5	>500	>140	≥11

Multiparameter optimization has also been explored for CNS drugs (Wager et al. 2010), because these drugs require a more narrowly defined chemical space. The individual scoring functions for six parameters, clog P, clog D, MW, TPSA, number of HBDs and pK_a for CNS drugs are presented in Fig. 9.7. All parameters are weighted equally with the maximum value for each parameter being 1 and the final score being a summation of the value of all 6 parameters.

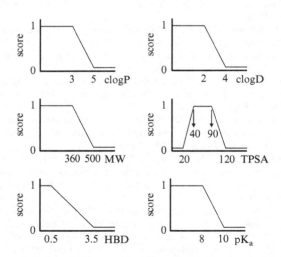

FIGURE 9.7. Example of scoring functions for clog P, clog D, molecular weight (MW), topological suface area (TPSA), hydrogen bond donors (HBDs) and pK_a to enable multiparameter optimization for central nervous system (CNS) drugs.

References

Chow MS (1996) Intravenous amiodarone: pharmacology, pharmacokinetics, and clinical use. Ann Pharmacother 30:637–643

Day GM, Cooper TG, Cruz-Cabeza AJ et al (2009) Significant progress in predicting the crystal structures of small organic molecules – a report on the fourth blind test. Acta Cryst B65:107–125

Doggrell SA (2001) Amiodarone – waxed and waned and waxed again. Expert Opin Pharacother 2:1877–1890

Ertl P, Rohde B, Selzer P (2000) Fast calculation of molecular polar surface area as sum of fragment-based contributions and its application to the prediction of drug transport properties. J Med Chem 43:3714–3717

Johnson TW, Dress KR, Edwards M (2009) Using the golden triangle to optimize clearance and oral absorption. Bioorg Med Chem Lett 19:5560–5564

Leeson PD, Davis AM (2004) Time-related differences in the physical property profiles of oral drugs. J Med Chem 47:6338–6348

Leeson PD, Springthorpe P (2007) The influence of drug-like concepts on decision-making in medicinal chemistry. Nat Rev Drug Discov 6:881–890

Lobell M, Hendrix M, Hinzen B et al (2006) In silico ADMET traffic lights as a tool for the prioritization of HTS hits. ChemMedChem 1:1229–1236

Lovering F, Bikker J, Humblet C (2009) Escape from flatland: increasing saturation as an approach to improving clinical success. J Med Chem 52:6752–6756

Morphy R (2006) The influence of target family and functional activity on the physicochemical properties of pre-clinical compounds. J Med Chem 49:2969–2978

Pajouhesh H, Lenz GR (2005) Medicinal chemical properties of successful central nervous system drugs. NeuroRx 2:541–553

Paolini GV, Shapland RHB, Van Hoorn WP et al (2006) Global mapping of pharmacological space. Nat Biotechnol 24:805–815

Price DA, Blagg J, Jones L et al (2009) Physicochemical drug properties associated with in vivo toxicological outcomes: a review. Expert Opin Drug Metab Toxicol 5:921–931

Ritchie TJ, Macdonald SJF (2009) The impact of aromatic ring count on compound developability – are too many aromatic rings a liability in drug design? Drug Disc Today 14:1011–1020

Ritchie TJ, Macdonald SJF, Young RJ, Pickett SD (2011) The impact of arumatic ring count on compound develop ability further indiunts by examining carbo- and hetero-arumatic and -Aliphatic ring types Drug Disc today 16:164–171

Segall M, Champness E, Obrezanova C et al (2009) Beyond profiling: using ADMET models to guide decisions. Chem Biodivers 6:2144–2151

Varma MVS, Obach RS, Rotter C et al (2010) Physicochemical space for optimum oral bioavailability: contribution of human intestinal absorption and first-pass elimination. J Med Chem 53:1098–1108

Vieth N, Siegel MG, Higgs RE et al (2004) Characteristic physical properties and structural fragments of marketed oral drugs. J Med Chem 47:224–232

Wager TT, Hou X, Verhoest PR et al (2010) Moving beyond rules: the development of a central nervous system multiparameter optimization (CNS MPO) approach to enable alignment of druglike properties. ACS Chem Neurosci 1:435–449

Wenlock MC, Austin RP, Barton P et al (2003) A comparison of physicochemical property profiles of development and marketed oral drugs. J Med Chem 46:1250–1256

Additional Reading

Kerns EH, Di L (2008) Drug-like properties: concepts, structure design and methods: from ADME to toxicity optimization. Academic Press, Amsterdam, The Netherlands

Mannhold R (2008) Molecular drug properties: measurement and prediction. Wiley-VCH, Weinheim, Germany

Van De Waterbeemd H, Testa B (2008) Drug bioavailability: estimation of solubility, permeability, absorption and bioavailability, 2nd edn. Wiley VCH, Weinhein, Germany

Chapter 10
In Silico ADME Tools

Abstract
The breath and predictive power of in silico ADME tools has increased rapidly during the last 10 years. The quality of many models is such that they can successfully influence decision making in drug discovery and development. In drug discovery they can influence decision related to synthesis of compounds and in development they can influence decisions to perform certain clinical trials or the design of the trial.

Contents

10.1 ABBREVIATIONS

CAT   Compartmental absorption and transit model
PBPK   Physiologically based pharmacokinetic model
P450   Cytochrome P450
PLS   Partial least squares
SAR   Structure–activity relationship
SVM   Support vector machine

10.2 BASIC CONCEPTS

In silico tools have become remarkably powerful and predictive in ADME sciences and are beginning to play a key role in drug discovery and development. They can be employed prior to synthesis

S.C. Khojasteh et al., *Drug Metabolism and Pharmacokinetics Quick Guide*, DOI 10.1007/978-1-4419-5629-3_10, © Springer Science+Business Media, LLC 2011

of compounds to improve the likelihood of identifying compounds with acceptable ADME properties. Moreover, if a poor ADME property is predicted with confidence, the in silico data can be used to reduce the number of compounds going through a particular assay. In silico models can also be used during lead optimization to help predict human pharmacokinetics. Finally, in silico models are being used in drug development to predict (1) the likelihood of encountering drug–drug interactions (such models can also influence the design of these drug–drug interaction studies), (2) formulation and physical form effects, (3) food effects, or (4) effects of different dosing regimens. In silico ADME models can simplistically be divided into two categories:

- Quantitative structure–activity (or property) relationship (SAR) models for specific in vitro ADME assays (such as metabolic stability in microsomes) based on a training set and a range of molecular descriptors describing the structure of the compounds in the training set.
- Physiologically based pharmacokinetic (PBPK) models reflecting an integrated system.

10.3 STRUCTURE-BASED MODELS

A range of software packages are available to predict basic properties such as pKa, log P, log D, and TPSA using the structure of the compound as the input. Nevertheless, some basic properties, such as solubility, are still remarkably hard to predict accurately. Specific ADME models are commercially available as well (see Table 10.1). However, the most successful ADME models tend to be built in-house because pharmaceutical companies have large data archives, and these data were acquired using the same assay format (Gao et al. 2008; Gleeson et al. 2007; Lee et al. 2007; Stoner et al. 2006). First, a large, structurally diverse training set is used to build the models. It is critical that the training set contains compounds covering the whole range of the ADME property under investigation. Inputs for the model are the structure of the compounds and the measured in vitro ADME parameter (e.g., metabolic stability in microsomes, plasma protein binding). A large number of molecular descriptors are calculated for each compound in the training set, ranging from simple parameters such as molecular weight, log P, log D_{pH}, and TPSA to much more complex parameters reflecting the electronics and/or three-dimensional nature of the compounds. Next, an analysis is performed to identify those descriptors that most strongly correlate with the measured parameters. Multiple modeling methods are available to facilitate the model building and optimization process. Models can be defined in most cases by their output: classification or numerical. The result of a classification method falls in a number of bins while for regression methods the output is numerical (although converting it to bins may be more appropriate to prevent over interpretation). The most common statistic methodologies used to build DMPK models are

- Regression methods
  - Partial least squares (PLS)
- Bayesian methods
- Supervised learning methods:
  - Decision trees (random forest)
  - Support vector machine (SVM)
- Neural networks

Finally, one or more models is built using those descriptors that correlate best with the measured parameter. (The number of descriptors should be kept limited to prevent over fitting.) To validate the models, a second, independent dataset is used. The calculated values of the parameter of interest are compared with the measured values for this validation set, and the most predictive model is selected. The whole process is illustrated in Fig. 10.1.

TABLE 10.1. Most common commercially available software to predict various ADME properties using built-in models or the ability to build new models

Software	Vendor
ADMET Predictor	Simulations Plus
ADME Suite	ACD Labs
Discovery Studio	Accelrys
KnowItAll	Bio-Rad
QikProp	Schrödinger
Sarchitect	Strand Life Sciences
StarDrop	Optibrium
VolSurf	Molecular Discovery

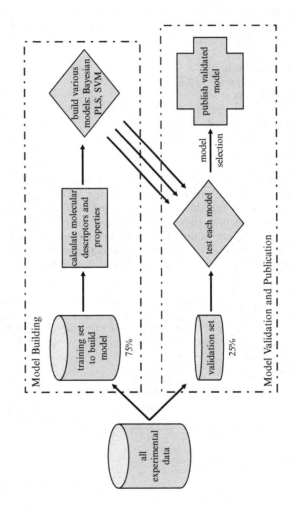

FIGURE 10.1. Flowchart depicting the building and validation of in silico ADME models. PLS = partial least squares; SVM = support vector machine.

Some models may have a continuous output, but frequently the output is based on categories (e.g., compounds are predicted to be metabolically stable, moderately stable, or labile). If the output is continuous, the correlation coefficient between the calculated and measured values for the validation set should be calculated to illustrate the predictive power of the model. If the output is based on categories (i.e., a classification model), the % false positives and negatives can be calculated. The output of each model should also include the *confidence in the prediction*, which is usually derived from (1) the structural similarity to compounds in the training set and (2) the number of nearest neighbors. Successful models have been built to predict ADME properties such as metabolic stability in microsomes or hepatocytes, cytochrome P450 (P450) competitive or time-dependent inhibition, permeability, plasma protein binding, and microsomal binding.

The following aspects should be considered when using in silico ADME models:

- It is possible to build global models based on a large, structurally diverse training set, but local models (based on data obtained for one particular project or chemotype) may be more predictive.
- It is important that the data used to build the model are obtained under identical or very similar experimental conditions.
- The whole range of the parameters should be covered by the training set and, preferably, to a similar extent.
- For a dynamic model, the calculated versus measured values should be monitored continuously.

10.3.1 Software to Predict Sites of Metabolism

The most widely used software packages to predict the most likely sites of metabolism are META, Meteor, MetabolExpert, StarDrop, and MetaSite (see also Table 10.2).

TABLE 10.2. Most common commercially available software to predict the most likely sites of metabolism

Software	Vendor
Meteor	Lhasa
META	Multicase
MetabolExpert	CompuDrug
MetaSite	Molecular Discovery
SMARTCyp	University of Copenhagen
StarDrop	Optibrium

1. META, Meteor, and MetabolExpert are rule-based systems built on a large compilation of biotransformation reactions presented in the literature. Predictions are based on substructure specific metabolism "rules" derived from the database, while ignoring the three-dimensional structure of the cytochrome P450 enzyme and the substrate.

2. The prediction of the site of metabolism in StarDrop is based on two factors: (1) the intrinsic reactivity of each potential site to oxidation by P450 enzymes and (2) the accessibility of the site of metabolism to the active oxy–heme species of P450, which is influenced by the orientation of the substrate in the active site and steric hindrance by nearby groups in the substrate. The reactivity component is the same for every P450 isoform as the reaction mechanism is believed to be similar for all isoforms of P450. However, the orientation and steric accessibility contributions vary between isoforms, reflecting the different binding pockets. The intrinsic reactivity is calculated by estimating the activation energy for the rate limiting step of the oxidation reaction using AM1, a semi-empirical quantum mechanical method. In the case of metabolism at an aliphatic carbon (leading to aliphatic hydroxylation, N- and O-dealkylation) the rate limiting step is hydrogen abstraction, and for aromatic sites formation of a tetrahedral intermediate between the substrate and the oxy–heme is rate limiting. The steric accessibility and orientation effects are estimated as contributions to the activation energy using models based on ligand structure that were trained using a large number of substrates for each P450 isoform (CYP2C9, CYP2D6, and CYP3A4). The final activation energies are then used to calculate the relative rates of product formation at the different sites of metabolism and, hence, the predicted regioselectivity.

3. MetaSite uses a slightly different computational procedure than StarDrop and considers the computed three-dimensional structure of the compound and GRID-based representations of P450 enzymes (1A2, 2C9, 2C19, 2D6, and 3A4). Descriptors are calculated for both, and the fingerprints of both are compared. The comparison provides two key parameters: (1) the accessibility of all molecular features of the drug in the active site of the P450 enzyme toward the heme group and (2) the reactivity of the molecular substructure (based on molecular orbital calculations and fragment recognition). Prediction of the site of metabolism is based on a probabilistic calculation taking both proximity

to the reactive oxygen species in the P450 binding pocket and reactivity into consideration (Cruciani et al. 2005). Although valuable and quite predictive, MetaSite provides only the site of metabolism, but not the metabolic pathway.

These in silico models have been successfully integrated in the metabolite identification process and can facilitate or speed up interpretation of data (while keeping in mind that the models will not be 100% predictive, especially if novel biotransformation pathways are involved). However, these models are usually of limited use in predicting the absolute importance of individual metabolic pathways, and StarDrop and MetaSite cannot predict non-P450 mediated metabolism.

10.4 PHYSIOLOGICALLY BASED PHARMACOKINETIC MODELS

Physiologically based pharmacokinetic (PBPK) models (Lavé et al. 2007) are more sophisticated than allometric approaches or simple in vitro–in vivo extrapolation for PK prediction. PBPK models are built around a wide range of parameters that describe the normal physiology of the human or animal body. These models are made up of multiple compartments, each representing a predefined tissue or organ, and are connected via blood or lymph flow, as is depicted in Fig. 10.2. Parameters included in these models are those related to human physiology, such as blood flow to organs, weight of organs, drug metabolizing enzymes in the liver and elsewhere in the body, and drug transporters in the body. Oral absorption is frequently modeled using augmented versions of the Compartmental Absorption and Transit (CAT) model. These models describe the various compartments of the intestinal tract and can include details such as:

- Rate constants for gastric emptying
- Intestinal compartmental transit time
- Local permeability of intestinal wall
- pH of each compartment
- Volume and surface area of each compartment
- Enterocytic blood flow

These models integrate the given data to determine the extent of dissolution, absorption, and metabolism (e.g., hepatic first pass, enterocyte metabolism) for each intestinal compartment. A large

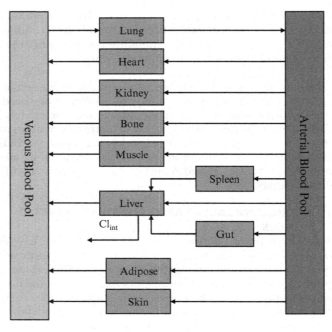

→ Represents blood flows and clearances

FIGURE 10.2. Physiological compartments incorporated in physiologically based pharmacokinetic (PBPK) models.

amount of data tends to be required for PBPK models. The input for PBPK models are parameters that specifically describe the compound of interest: pKa, lipophilicity, solubility in various media, in vitro permeability, in vitro metabolic stability, and P450 inhibition, among others. In addition, the dose, dosing route, and dosing regimen need to be specified. Commercially available PBPK models are listed in Table 10.3.

TABLE 10.3. Most common commercially available PBPK models

Software	Vendor
Cloe PK	Cyprotex
GastroPlus	Simulations Plus
PK-Sim	Bayer
Simcyp	Simcyp Consortium

The output of the PBPK model is a concentration–time profile. For some models, it is possible to provide a population range instead of an average profile. PBPK models are quite powerful, and they are extensively used for predicting and understanding the factors influencing the following phenomena:

- Human and animal pharmacokinetics
- Food effects
- Formulation effects
- Dosing regimen effects
- Drug–drug interactions via competitive and time-dependent P450 inhibition and P450 induction

References
Cruciani G, Carosati E, De Boeck B et al (2005) MetaSite: Understanding metabolism in human cytochromes from the perspective of the chemist. J Med Chem 48:6970–6979

Gao H, Yao L, Mathieu HW et al (2008) In silico modeling of nonspecific binding to human liver microsomes. Drug Metab Dispos 36:2130–2135

Gleeson MP, Davis AM, Chohan KK (2007) Generation of in–silico cytochrome P450 1A2, 2C9, 2C19, 2D6, and 3A4 inhibition QSAR models. J Comput Aided Mol Des 21:559–573

Lavé T, Parrott N, Grimm HP et al (2007) Challenges and opportunities with modeling and simulation in drug discovery and drug development. Xenobiotica 37:1295–1310

Lee PH, Cucurull-Sanchez L, Lu J et al (2007) Development of in silico models for human liver microsomal stability. J Comput Aided Mol Des 21:665–673

Stoner CL, Troutman M, Gao H et al (2006) Moving in silico screening into practice: A minimalist approach to guide permeability screening. Lett Drug Des Discov 3:575–581

Additional Reading
Espié P, Tytgat D, Sargentini–Maier M-L et al (2009) Physiologically based pharmacokinetics (PBPK). Drug Metab Rev 41:391–407

Hou T, Wang J (2008) Structure–ADME relationship: still a long way to go. Expert Opin Drug Metab Toxicol 4:759–770

Kharkar PS (2010) Two-dimensional (2D) in silico models for absorption, distribution, metabolism, excretion and toxicity (ADME/T) in drug discovery. Curr Top Med Chem 10:116–126

Chapter 11
Approved Drugs

Abstract
Here we represent resources available, including FDA web sites, for information on the drug approval process and Critical Path Initiatives. We include information on the cost of clinical drug development, approved drugs since 1995, the largest pharmaceutical companies based on pharmaceutical sales, the top selling drugs in 2009, and the approved drugs in the first half of 2010.

Contents

11.1 ABBREVIATIONS
CDER Center for Drug Evaluation and Research
CPI Critical Path Initiative
IND Investigational new drug
NDA New drug application

11.2 HOW DRUGS ARE APPROVED BY THE FDA
The Center for Drug Evaluation and Research (CDER) regulates the process for testing drugs in humans. Before any drug is tested in clinical studies in the United States, the CDER must approve an investigational new drug (IND) application. For small molecules, a new drug application (NDA) must be approved after phase III

S.C. Khojasteh et al., *Drug Metabolism and Pharmacokinetics Quick Guide*, DOI 10.1007/978-1-4419-5629-3_11,
© Springer Science+Business Media, LLC 2011

clinical studies in order to market the drug (Tables 11.1–11.6). To learn more go to:

How Drugs are Developed and Approved (2010) FDA. http://www.fda.gov/Drugs/DevelopmentApprovalProcess/HowDrugsare-DevelopedandApproved/default.htm. Accessed 3 September 2010

The FDA's *Critical Path Initiative* (CPI) is a strategy for expediting and improving the drug development process.

To learn more go to:

Critical Path Initiative (2010) FDA. http://www.fda.gov/Science-Research/SpecialTopics/CriticalPathInitiative/default.htm. Accessed 3 September 2010

TABLE 11.1. The cost of drug development (DiMasi et al. 2003)

Clinical phase	Mean (SD) in millions of dollars
I	15.2 (14.3)
II	41.7 (30.2)
III	115.2 (95.0)

TABLE 11.2. FDA-approved drugs since 1995 based on therapeutic area

Therapeutic area	1995–1999	2000–2004	2005–2009
Cardiology/vascular diseases	51	23	15
Dental/maxillofacial surgery	5	2	0
Dermatology/plastic surgery	31	15	8
Endocrinology	126	42	14
Gastroenterology	23	27	0
Hematology	15	12	17
Immunology/infectious diseases	64	40	38
Musculoskeletal	25	17	16
Nephrology/urology	25	22	12
Neurology	49	28	24
Obstetrics/gynecology	56	29	6
Oncology	57	33	26
Ophthalmology	17	9	10
Otolaryngology	7	3	5
Pediatrics/neonatology	38	19	24
Pharmacology/toxicology	11	3	3
Psychiatry/psychology	27	16	9
Pulmonary/respiratory diseases	54	21	8
Rheumatology	13	10	8
Trauma/emergency medicine	2	0	2
Total	696	371	245

FDA-Approved Drugs by Therapeutic Area (2010) Centerwatch. http://www.centerwatch.com/drug-information/fda-approvals/drug-areas.aspx. Accessed 3 September 2010

TABLE 11.3. The total revenue from the pharmaceutical, biological and generic drug markets and the top three therapeutic areas in 2009 in billions of dollars

Drug market	Revenue ($ billions)	Revenue from the top three therapeutic areas ($ billions)
Pharmaceutical	810	Central nervous system → 125 Cardiovascular → 110 Oncology → 75
Biological	130	Monoclonal antibodies →40 Vaccines →25 TNF inhibitors → 22
Generics	90	

Top Ten/Twenty Best Selling Drugs 2009 (2009) Knol. http://knol.google.com/k/top-ten-twenty-best-selling-drugs-2009. Accessed 3 September 2010

TABLE 11.4. The top 12 largest pharmaceutical companies based on total revenue from pharmaceutical sales in 2009 (source annual reports)

Rank	Company	Country	Total revenue ($ billions)
1	Pfizer	USA	45.448
2	Sanofi-Aventis	France	36.131
3	GlaxoSmithKline	UK	35.127
4	Roche	Switzerland	34.522
5	AstraZeneca	UK	31.905
6	Novartis	Switzerland	28.538
7	Merck & Co.	USA	25.236
8	Johnson & Johnson	USA	22.520
9	Eli Lilly	USA	21.175
10	Abbott Laboratories	USA	16.486
11	Bristol-Myers Squibb	USA	15.272
12	Bayer HealthCare	Germany	12.375

TABLE 11.5. The top selling drugs in 2009

Generic name (brand name)	Company	Indication	2009 sales ($ billions)
Atorvastatin (Lipitor)	Pfizer, Astellas	High cholesterol	12.45
Clopidogrel (Plavix)	Bristol-Myers Squibb, Sanofi Aventis	Acute coronary syndrome (reduction of thrombotic events)	9.29
Etanercept (Enbrel)	Amgen, Pfizer, Takeda	Rheumatoid arthritis, juvenile rheumatoid arthritis, psoriasis, psoriatic arthritis, ankylosing spondylitis	8.0
Fluticasone, Salmeterol (Advair)	GlaxoSmithKline	Asthma	7.76
Infliximab (Remicade)	Johnson & Johnson, Merck, Mitsubishi Tanabe	Rheumatoid arthritis, psoriasis, psoriatic arthritis, ankylosing spondylitis, ulcerative colitis, Crohn's disease	6.91
Valsartan (Diovan)	Novartis	Hypertension	6.01
Bevacizumab (Avastin)	Roche	Colon cancer	5.92
Rituximab (Rituxan)	Roche	Non-Hodgkin's lymphoma, rheumatoid arthritis	5.80
Aripiprazole (Abilify)	Otsuka, Bristol-Myers Squibb	Schizophrenia	5.6
Adalimumab (Humira)	Abbott Laboratories	Rheumatoid arthritis, psoriasis, juvenile idiopathic arthritis, psoriatic arthritis, ankylosing spondylitis, Crohn's disease	5.49
Trastuzumab (Herceptin)	Roche	Breast Cancer	5.02
Esomeprazole (Nexium)	Astra-Zeneca	Ulcers	4.95
Olanzapine (Zyprexa)	Eli Lilly	Schizophrenia	4.91
Quetiapine (Seroquel)	AstraZeneca Astellas	Schizophrenia	4.89
Rosuvastatin (Crestor)	AstraZeneca, Shionoggi	High cholesterol	4.74

Continued

TABLE 11.5 *Continued*

Generic name (brand name)	Company	Indication	2009 sales ($ billions)
Montelukast (Singulair)	Merck	Asthma	4.66
Venlafaxine (Effexor)	Pfizer	Depression	4.3
Insulin glargine (Lantus)	Sanofi Aventis	Diabetes	4.22
Enoxaparin (Lovenox)	Sanofi Aventis	Deep vein thrombosis	4.17
Pioglitazone (Actos)	Takeda	Diabetes	4.11

TABLE 11.6.   Drugs approved in the first half of 2010 (January–August)

Generic name (brand name)	Company	Indication	Approval date
Tocilizumab (Actemra)	Roche	Rheumatoid arthritis	January 2010
Liraglutide (Victoza)	Novo Nordisk	Type 2 diabetes mellitus	January 2010
Dalfampridine (Ampyra)	Acorda	Improved walking in patients with multiple sclerosis	January 2010
Aztreonam (Cayston)	Gilead Sciences	Cystic fibrosis	February 2010
Meningitis vaccine (Menveo)	Novartis	Prevention of invasive meningococcal disease	February 2010
Trazodone hydrochloride (Oleptro)	Labopharm	Depression	February 2010
Pneumococcal 13-valent conjugate vaccine (Prevnar 13)	Wyeth	Prevention of invasive disease caused by Streptococcus pneumoniae	February 2010
Collagenase clostridium histolyticum (Xiaflex)	Auxilium Pharmaceuticals	Dupuytren's contracture	February 2010
Velaglucerase alfa (Vpriv)	Shire	Type 1 Gaucher disease	March 2010
Imiquimod (Zyclara)	Graceway	Actinic keratoses of the face and scalp	March 2010
OnabotulinumtoxinA (Botox)	Allergen	Upper limb spasticity	March 2010
Carglumic acid (Carbaglu)	Recordati	Hyperammonemia	March 2010
Hydromorphone hydrochloride (Exalgo extended release)	Alza	Pain	March 2010
Doxepin (Silenor)	Somaxon Pharma	Insomnia	March 2010
Rifaximin (Xifaxan)	Salix	Hepatic encephalopathy	March 2010
Miconazole (Oravig)	Strativa Pharmaceuticals	Oropharyngeal candidiasis	April 2010
Pancrelipase (Pancreaze)	Johnson & Johnson	Exocrine pancreatic insufficiency	April 2010
Naproxen + esomeprazole (Vimovo)	AstraZeneca	Arthritis	April 2010

Drug	Company	Indication	Date
Everolimus (Zortress)	Novartis	Prevention of kidney rejection after transplantation	May 2010
Gatifloxacin ophthalmic solution (Zymaxid)	Allergen	Bacterial conjunctivitis	May 2010
Ketorolac tromethamine (Sprix)	Roxro Pharma	Pain	May 2010
Estradiol valerate + dienogest (Natazia)	Bayer	Prevention of contraception	May 2010
Sipuleucel-T (Provenge)	Dendreon	Hormone refractory prostate cancer	May 2010
Mometasone furoate + formoterol fumarate dihydrate (Dulera)	Merck	Asthma	June 2010
Denosumab (Prolia)	Amgen	Osteoporosis	June 2010
Cabazitaxel (Jevtana)	sanofi aventis	Prostrate cancer	June 2010
Olmesartan medoxomil + amlodipine + hydrochlorothiazide (Tribenzor)	Daiichi Sayko	Hypertension	July 2010
Clindamycin phosphate + tretinoin (Veltin)	Stiefel	Acne vulgaris	July 2010
Ondansetron oral soluble film (Zuplenz)	Strativa Pharmaceuticals	Chemotherapy and radiotherapy induced nausea and vomiting	July 2010
IncobotulinumtoxinA (Xeomin)	Merz Pharmaceutical	Cervical dystonia and blepharospasm	July 2010
Glycopyrrolate (Cuvposa)	Shionogi	Chronic severe drooling in pediatrics	July 2010
Ulipristal acetate (Ella)	HRA Pharma	Prevention of contraception	August 2010

Lists of drugs approved in previous years can be found at:

FDA-Approved Drugs by Year (2010) Centerwatch. http://www.centerwatch.com/drug-information/fda-approvals/. Accessed 3 September 2010

References

DiMasi JA, Hansen RW, Grabowski HG (2003) The price of innovation: new estimates of drug development costs. J Health Econ 22:151–185

Chapter 12
Chemical Nomenclature

Abstract
Here we present information on the naming convention of various organic chemicals and common moieties used by medicinal chemists. We also included figures of typical five-membered, six-membered and bicyclic heterocyclic rings.

Contents

12.1 GENERAL NOMENCLATURE FOR ORGANIC COMPOUNDS

General prefix for a carbon chain of any length: alk-
  General suffix for a functional group moiety: -yl

TABLE 12.1. Prefixes based on length of carbon chain				
	C1: meth-	C4: but-	C7: hept-	C10: dec-
	C2: eth-	C5: pent-	C8: oct-	
	C3: prop-	C6: hex-	C9: non-	

TABLE 12.2. Suffixes based on carbon saturation		
	Single bond (saturated)	-ane
	Double bond	-ene
	Triple bond	-yne

S.C. Khojasteh et al., *Drug Metabolism and Pharmacokinetics Quick Guide*, DOI 10.1007/978-1-4419-5629-3_12, © Springer Science+Business Media, LLC 2011

TABLE 12.3. Suffixes and prefixes for common functional groups

Functional group	Suffix	Prefix
Acyl	-oyl	Acyl-
Aldehyde	-al	None
Alcohol	-ol	Hydroxyl-
Amide	-oic	None
Carboxylic acid	-oic acid	None
Carboxylic acid salt or ester	-oate	None
Ketone	-one	Keto- or oxo-
Nitrile	-nitrile	Cyano-

TABLE 12.4. Prefixes for heteroatoms

Element	Valence	Prefix
Nitrogen (N)	3	Aza-
Oxygen (O)	2	Oxa-
Sulfur (S)	2	Thia-
Phosphorous (P)	3	Phospha-

TABLE 12.5. Common suffixes for nitrogen-containing and non-nitrogen-containing heterocycles

Ring size	Saturated		Unsaturated	
	with N	without N	with N	without N
3	-iridine	-irane	-irine	-irene
4	-etidine	-etane	-ete	-ete
5	-olidine	-olane	-ole	-ole
6	-inine	-inane (-ane)	-ine	-ine

Note that if the ring contains some degree of unsaturation, the prefixes dihydro- or tetrahydro- are added accordingly.

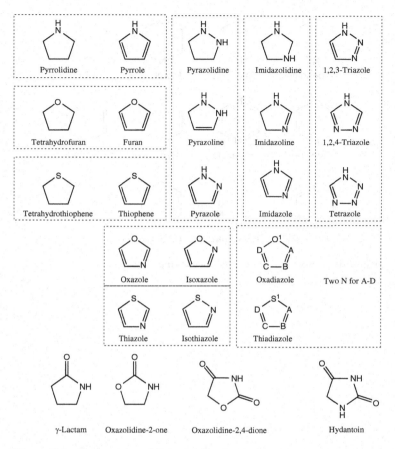

FIGURE 12.1. Five-membered heterocyclic rings.

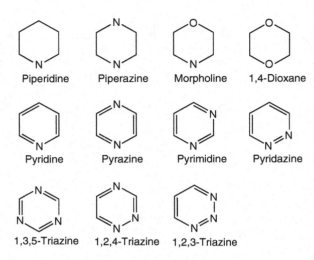

FIGURE 12.2. Six-membered heterocyclic rings.

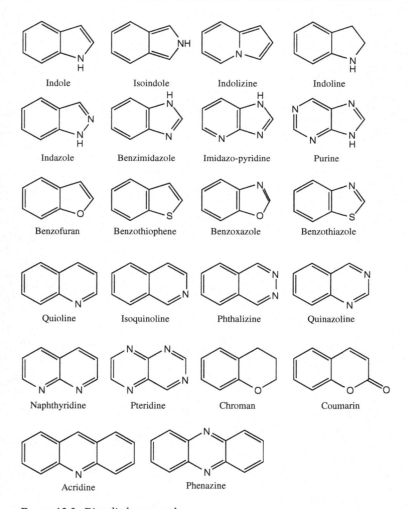

FIGURE 12.3. Bicyclic heterocycles.

Index